'Solving the climate crisis requires changes from the global to garden level and Mia provides a simple and savvy handbook on how to get started in your own home.'

Richie Merzian, Director Climate & Energy Program
The Australia Institute

'Mia's book is the everyday person's antidote to the climate crisis. Full of practical and enticing examples, it is a great guide to changes we can all make to create a better climate and save money and be healthier at the same time. It shows how everyone can be a part of the climate solution.'

Professor Mark Howden, Director of the Institute for Climate, Energy and Disaster Solutions, Australian National University. Major contributing author to Intergovernmental Panel on Climate Change (IPCC) Assessment reports two, three, four, five and six, sharing the 2007 Nobel Peace Prize with other IPCC participants and Al Gore.

'Happy Planet Living is for every person who wants to know how they can make a difference to our planet in their own life. Mia guides the reader to understand the choices they have to help shape a better future for us all.'

Dr Tara Shine, CEO of Change by Degrees, TV presenter and author of *How to Save Your Planet One Object at a Time*

'Mia's passion for making sustainability simple is infectious. What I love about Mia is how she is inspiring and practical, yet not at all judgmental. Her approach is a breath of fresh air in an era when many people are experiencing eco-anxiety.'

Serina Bird, author of *the Joyful Frugalista* and *the Joyful Startup Guide*

HAPPY PLANET LIVING

Simple Ways to Live A Climate Positive Lifestyle & Make A Big Difference

MIA SWAINSON

Published by Wilkinson Publishing Pty Ltd
ACN 006 042 173
Level 6, 174 Collins Street, Melbourne, VIC 3000, Australia
Ph: +61 3 9654 5446
enquiries@wilkinsonpublishing.com.au
www.wilkinsonpublishing.com.au

ISBN: 9781922810168
eISBN: 9781922810946
A catalogue record for this book is available from the National Library of Australia.

Photos on pages 158, 177, 179, 186, 191, 198 and 207 by Nick Taylor.
Photos on pages 37, 52, 53, 68 (bottom), 73, 85, 133, 151, 192, 209 courtesy of istock.com.
All other photos by Dion Georgopoulos
Design by Tango Media

Printed in Australia by Ligare Book Printers

The paper this book is printed on is in accordance with the standards of the Forest Stewardship Council®. The FSC® promotes environmentally responsible, socially beneficial and economically viable management of the world's forests.

Follow Wilkinson Publishing on social media.

 WilkinsonPublishing

 wilkinsonpublishinghouse

 WPBooks

For Ashwyn, Xavier and Tasman

Contents

Author's note

I have always wanted to create a better future for people and our planet. This was my driver for studying environmental engineering at university, then for working in international development and government. Somewhere along the way I realised that individual choices made by everyday people – in where they spend money, how they travel and how they think about the environment - underpin the relationship between people and our planet. Right now our planet's in crisis. We don't have time to wait for governments or big business to act. We have the power to create change right now, with our everyday, ordinary lives.

I wrote Happy Planet Living because I wanted people to know that making a difference is remarkably simple. We don't have to choose between life in an off-grid commune and modern conveniences. My book is a guide to the middle way. Make changes one step at a time… and have some fun along the way.

INTRODUCTION

Modern life is increasingly disconnected from nature and planet earth. Fruit and vegetables come from the shops, not our gardens. We eat the same pre-packaged food, in every season. We buy things because they are cheap, not because they will last.

If human civilisation is to survive, we need to regenerate and strike a balance with nature. Our lives and our economies exist because of the natural environment on our planet. We cannot survive without a thriving natural environment.

There is hope. It is possible for us to restore nature, live climate positive lives, in ways that also make us financially better off to enjoy the benefits of our modern world. The small choices that we make every day can make a big difference, to the planet and to our lives.

Around the world, if every person refused single use plastic drinking bottles we could save one million plastic bottles every minute. If every person refused single use plastic bags, we could save 5 trillion single-use plastic bags each year. The lives of thousands of marine animals, some endangered, could be saved. If every family in the United States chose to be carbon neutral, then the country's greenhouse gas emissions would be reduced by 80%. The choices we make in our homes and local communities are profound for the world.

There are simple pleasures to be had with a sustainable life. Growing seedlings in a garden, big or small, creates excitement, anticipation and reward. Making jams, chutneys and flavoured oils can create a unique dinner experience. Choosing to buy less stuff creates more time and boosts your bank balance.

Living in balance with nature is just as achievable for people who live in city apartments as it is for people living in an off-grid community. Growing your own herbs, salad and microgreens is possible in small spaces. A vegetarian diet, ethically investing your retirement pension, upcycling clothes or furniture all work in city apartment living.

Sustainable and regenerative living is not new, it is just a forgotten art. Most of our great grandparents knew how to make their own jam, repair their own clothes, and buy stuff that lasted. Growing our own food connects us to the ecosystem in which we live and was commonplace until the 1950s. Making yoghurt is an ancient art, stretching back more than 5000 years in Central Asia.

Enjoy the pleasures to be had from even the smallest taste of sustainable living. Tackle change in your life by taking one step at a time to create a happy planet.

Feel good about what you can do, not guilty about what you can't. The key to success is making the change, one step at a time. Choose the changes that are right for you, at this time in your life. Over time, create your own paradise in urban sustainable living and a happier planet.

Mia Swainson

HOME WITH A SMALL FOOTPRINT

FASTEST WAY TO A HAPPY PLANET

Time is of the essence. Our planet's climate, biodiversity and natural resources are in desperate need of regeneration. We can all make a difference through the choices we make each day. Here are the most impactful places to start. These are seven fast, simple ways for us to play our part in creating a happy planet.

1. Fly less

There's no way to sugar-coat the impact of flights on carbon emissions, especially long-haul flights. Flying from London to New York and back is about 2 tonnes of emissions. That's almost one third of the average emissions per individual who lives in the United Kingdom. Fly only when necessary. Consider holidays that are closer to home. Swap flights for video conference calls at work.

2. Buy things that you *really* need

Shiny new things are just so alluring! The latest computer, phone or Fitbit are fun. There are gadgets for your kitchen, laundry, shed and more. Then, there's fashion. New colours mean new tops, pants and matching shoes. Choose wisely. New things become old quickly. Many objects spend less time in our home than they spent getting manufactured and shipped for your buying pleasure. Try extending the life of things you already own, buying second hand or swapping on your local buy nothing group. Take a buy nothing new challenge.

3. Choose an ethical retirement pension

Retirement pensions are the single largest asset pool in the global market. Let your money express your values. Choose an ethical retirement pension provider that expresses your values of climate solutions and sustainable development. The good news is that you don't need to compromise your savings, most ethical pension funds perform slightly better than the average fund.

4. Recycle... everything

Before you put an item in the bin to go to landfill, ask yourself... can this be recycled? If you're not sure, check out the advice from your local council. We have the technology to recycle, but many people lack the willpower to sort their recycling properly. By recycling, you'll be creating a new life for the items in your home and reducing the need for mining and pre-production.

5. Choose renewable electricity

Choosing renewable electricity can be as simple as contacting your provider and asking to switch. They'll be buying your renewable energy in an electricity market — through the normal electricity grid. This drives up demand for new renewable electricity projects. Alternatively, install your own solar panels and battery.

6. Eat less meat

Food production is responsible for a quarter of total greenhouse gas emissions worldwide, with meat, in particular beef cattle and sheep, as the largest contributors. Beef and sheep produce methane when they fart, they're also responsible for deforestation to create agricultural land. If every person in Europe decided to be vegetarian tomorrow, their emissions from food would be sliced in half. Fall in love with tasty Indonesian Gado Gado, Middle Eastern Falafel burgers and Mexican bean tacos.

7. Use the avoid, reuse and reduce compass to guide household decisions

Need something new? Think again. Can you live without it, or avoid the purchase. Do you already have a mountain bike or kitchen gadget that's good enough. If you really do need something, can you reuse some or all of something that you already have. Can you reduce the number or size of the thing that you need. For example, do you really need a new home, or would an extension be sufficient? Do you need a new car, or would a trailer enable you to keep your existing car. Can you live with less, so the planet can have more?

8. Grow some of your food

Connect to the climate and landscape near your home by growing some of your own food. Start with herbs or leafy greens in pots. Use the food that you grow. Extend your range as you get to know and enjoy being outside.

CARBON POSITIVE HOME

Global warming looms like a tribe of rampaging giants. It is a future with more drought, bushfires, smoke haze and rising sea levels. How does one person stand up against the tide to make a change when the problem is so big?

Just as climate change is a challenge that was created by the actions of people, it is a challenge that people will need to solve. We can be optimistic about the future. Around the world, people are standing together to make a difference on climate change. Greta Thunberg began leading the movement for change when she was still a high school student. More than 1 Million women have joined together in a campaign to build a lifestyle revolution to save the planet. Going carbon positive is a change that we need to see in people's lives to regenerate our planet. Together we can turn the tide of climate changing, rampaging giants. Living carbon positive is just a few steps away.

Step 1. Know your carbon footprint
Step 2. Decide what you're going to change (or not)
Step 3. Offset the difference

Step 1. Know your carbon footprint

The choices we make every day have a profound effect on our personal carbon footprint. Do you fly to your holiday destination or choose somewhere that's accessible by car? Have you got solar panels or do you pay a premium for renewable energy through the electricity grid? Do you like to buy new stuff… clothes, shoes, computers or furniture, or do you search out second hand? Do you eat local tomatoes all year, or only local tomatoes when they're in season? Should you be vegetarian? So many choices.

Here are three typical annual carbon footprints built using the United Nations Climate Change Lifestyle calculator. These are from households around the world but the names are fictional.

Ash lives alone in London, takes the underground train to and from work, doesn't own a car and takes expensive holidays by train (not flight). Ash is a typical consumer, buying quite a few new things each year. His electricity is directly from the grid and his diet is vegetarian.

Ash's carbon footprint is quite low because he hasn't taken any flights. Being vegetarian also makes a difference, the carbon contribution of his diet is about half of the carbon contribution that it might have been if he had a typical meat-eating diet. To reduce emissions even further, Ash might choose to source all of his home's energy from a renewable source, or to reduce his consumption of new products.

Miranda lives in Australia with her partner and two children. They have a medium-sized house, a car and she has taken one short flight and one long flight for holidays. Miranda catches the bus to and from work each day and has a hybrid car. She is a normal consumer, buying new clothes, shoes and small whitegoods this year. She has a 'normal' diet that includes meat.

Miranda's carbon footprint is quite high, driven mostly by the flights she's been taking as well as her her consumption and her diet. The fastest way for Miranda to reduce her footprint is to take less flights.

Rosie is elderly and lives in America with her partner in an average sized home. Rosie visits her grandchildren once each year, on a short flight each time. She eats meat and is a frugal shopper, only buying the things she needs and buying other things second hand.

Rosie's main source of emissions is from her home. The simplest way for her to reduce emissions is by choosing renewable energy through her existing energy provider or to install solar panels and a battery.

Navigating your carbon footprint can feel like you're navigating a lumpy minefield. Until every item that we buy has a carbon price attached, the simplest way to know your personal carbon footprint is to use an online calculator to do the heavy lifting.

The United Nations Climate Change Lifestyle Calculator is easy to use and builds from a sophisticated international data set. It takes about 4 minutes to complete an initial assessment, with more detailed assessment available. The Calculator helps you to understand your carbon footprint through the categories of home, transport, shopping and food. Other good carbon footprint calculators include the Global Footprint Calculator and Carbon Positive Australia's Calculator.

Step 2. Decide what you're going to change (or not)

Home

Electricity

Electricity can be provided carbon neutral in so many ways. The easiest option is to contact your electricity provider and ask for 'green power'. That means that you pay the additional cost for your electricity provider to source renewable energy through the grid. Most providers will simply buy *renewable energy credits* on the renewable energy market, stimulating even greater demand for renewable energy.

Another easy way to supply your home with renewable energy is to install solar panels, selling excess energy into the grid or storing excess energy in your own battery. Most solar panel installations pay off in 5–8 years, well before the panel's warranty period ends.

Gas

Typically used for heating water, heating your home and cooking. Each has alternatives that are carbon neutral — if your electricity is from a carbon neutral source. Consider improving efficiency in the way that you heat your home by adding insulation in the roof or sealing drafts.

Waste

Greenhouse gasses are created by food, garden waste and wood breaking down inside a landfill. The simple answer? Don't put these items into landfill. Many local councils offer organic waste collection options. If they don't there are options to suit any sized home, including home or community compost, a worm farm, and a bokashi bucket. There's more on how to take organic waste out of your bin to landfill in the zero waste section (page 46).

Recycling paper, glass and metal is another great way to reduce your greenhouse gas footprint. By recycling, you're reducing the energy needed to produce new products.

Transport

Vehicle

Does every adult in your family need their own car? Does every journey need to be by car? Does your car need to be run on fossil fuels?

Where you live and how you structure your life matters. Consider public transport, bikes and walking as alternatives. Consider owning just one car per family, rather than two or three.

Air travel

Do you need to take every trip by air? Are there holiday options closer to home?

Reducing the footprint of air travel is about avoiding and reducing the number and size of journeys by air. Reducing air travel is an easy way to reduce your carbon footprint.

Shopping

Every time we buy a new computer, new clothing or new appliance, there's been energy — usually resulting in carbon emissions — involved in the production and shipping of that item. Second hand items are usually locally sourced, so less shipping and there's no new emissions involved in creating the product. Take a buy nothing new challenge to see just how much stuff you didn't really need in the first place or could source second hand.

Food

Food makes a bigger contribution to your household's carbon emissions than most people realise. According to the UN Food and Agricultural Organisation, 14.5% of all human emissions are from livestock, most of this is from meat production. The greenhouse gas impact of different foods is surprising. (See chart on page 39)

We can see here that meat is the big contributor to food carbon. So, consider eating less meat or becoming vegetarian. If you do eat meat, consider chicken before anything else. Chicken is good because it doesn't require much grain to grow the meat, and because chickens don't fart quite so much as the other animals. Surprisingly, dairy has a bigger carbon impact than chicken... owing to both the transport of feed and the farting of cows. Consider cutting back on your dairy.

The other significant contributors to food carbon are the fossil fuels used to transport food and those used to heat greenhouses to grow out of season. Join the world's eat local movement and find ways to eat fresh, seasonal produce you're your local farmer's market.

Step 3. Offset the difference and make it positive

There's quite a bit of choice in the offsets market. The international gold standard certification body, endorsed by the United Nations for the offset activities having the highest level of environmental integrity as well as other benefits, like women's empowerment or improved biodiversity. Most of the certified offset projects are in developing countries and you can choose whether your offsets are from hydroelectricity, wind farms or biodiverse forests.

There are offset projects everywhere around the world. Choose an offset project that connects most deeply for you.

SUMMER-PROOF HOME

Here is how to summer-proof your home to create a cool sanctuary, every hot summer day. There are ideas to suit every budget and every type of home.

1. Shut up your home in the middle of the day and open it up at night. Let the lower overnight temperatures do the cooling for you. Benefits are greatest if you open doors and windows to allow a cross breeze. No breeze? Create one with a strategically placed fan that draws in the cool night air.

2. Turn down the heat inside your home. Slow roast in the oven for dinner in summer? No thanks. Try dinner cooked on a BBQ or the microwave to reduce the heat that you add into your home throughout the day.

3. Shade windows from the midday heat, especially on the western side of your home. Consider planting a pergola or deciduous shade trees along the western side of your home. Wisteria is a great vine for growing on a pergola or shade 'wall'. It's fast growing, impossible to kill, has lovely purple flowers in the spring time and drops its leaves in winter to let in the sunshine. Blinds, awnings or shade cloth are another option for shading windows.

4. Cool down a concrete patio or driveway with large pot plants. Lemon, lime and mandarin trees grow well on a deck or patio. They love the heat, as long as they're well-watered.

5. Seal gaps as they can be responsible for 15% of the heat that enters your home. It's cheap to seal them for most homes and there's no special building experience necessary. Ask how at your local hardware store or check out the online clips posted by Lish Fejer, the GreenItYourself expert.

6. Insulate your home. Think about insulating the ceiling, walls and floor. As most of the heat gained into your home in the summer is from the roof, focus on this first. For summer cooling, walls are the next priority for insulation, followed by the floor. Insulation is great because it has an impact for both winter and summer.

7. Keep your body cool. Think about the choices you'd make about your day if you lived in the tropics. Exercise in the early mornings, not in the middle of the day. If you're hot, take a short shower and feel refreshed. Take life at a slower pace.

WINTER-PROOF YOUR HOME

A cosy fireplace, a cup of hot tea and a good book are just perfect on a cold winter evening. They warm both our hearts and bodies at a time when morning frost decorates our landscape and temperature dominates the conversation.

If you're new to a cold climate, the first winter is the almost always the hardest. A warm home is the only way to survive. So, crank up the heating and walk around in a T-shirt? Oh no! That wouldn't work for our planet… or your heating bill. For most people, heating makes up the largest portion of household energy use. Ironically, this makes heating in cold climates the biggest global warming contributor from your activities at home.

The good news is that there's plenty that you can do to stay warm in winter, reduce your energy use and love your planet – all at the same time.

1. Start with you

Yes, really. Start by keeping your body warm as your first heating choice. Clothing layers are your friend – especially those gorgeous merino wool leggings, singlets and tops. Wear them underneath regular clothes throughout winter. Put on a jumper, rather than the home heater. For every degree that you dial down on your heater, that's about 10% off your total energy heating costs.

Another way to stay warm is to drink a cup of hot tea. Stay with me here, even if this sounds a little crazy. Patagonia, in the far south of South America, is absolutely freezing in the winter time. Most of the locals spend their day — yes, whole day — sipping enormous cups of *mate*, a type of herbal tea. Typically people carry it around with them, from home to work and back again.

Settling in to watch a movie on a cold night? Have a blanket or two handy in your living room and throw one over your knees if you start to get cold. Grab a hot water bottle or wheat bag to keep your toes warm.

2. Target your heating

Does the whole home need to be 20 degrees, including the bedrooms? Only if you hang out in the bedroom. Find a way to heat the space that you need, and not an inch more. You might need to install an extra door, or cover some of your heating vents. Be creative about the solutions in your home.

Get the timing right for your heating. Have a warm house when you're there and switch off the heating when you don't need it. Most central heating can be programmed. Some electric bar heaters have a built-in timer and for those that don't, you can purchase a separate timer for your heater's electricity plug. These days you can even purchase a *smart power socket* that lets you monitor power usage in real time and turn on and off appliances remotely using an app on your smartphone.

3. Keep your heat inside your home

Insulation is the key — in the roof, the walls and below your floor. In Canada and northern Europe homes are routinely insulated. Unfortunately for many other parts of the world, true heavy-duty insulation is a new phenomenon.

Next on the hit list are cracks and poor sealing doors that create currents of cold air. Want to know if this is a problem for you? Gently move your hand along the seal around your doors and windows to feel for that cold air current. You might be surprised. Sealing up cracks is simple and cheap.

Finally, if you live in a cold climate, consider double glazing all of the windows in your home.

Taking it one step at a time, you'll be warm on frosty mornings and kind to the planet at the same time.

Seal the gap between your front door and floor with a simple rubber and metal strip.

Seal the gap between your door and the wall with a foam strip.

LESS IS MORE AT HOME

Big houses, big cars and new shiny things. We love progress. It wasn't so long ago that the average family home had two rooms, no electricity and travel was by horse and cart. Wow. Life was tough back then.

Right now things are pretty comfortable. But, do we really need everything to be shiny, new and big to live comfortably? When we're building a new house, we find it hard to choose between two or three bathrooms. Is that second car really needed? We can always justify buying the latest release computer or phone... somehow?

Small houses cost less to build, use less energy for heating and cooling, need fewer furnishings and take less time to maintain. Australians and Americans have some of the biggest homes in the world, according to shrink my footprint, with an average of 77–89 m2/ person. This is almost three times the average floor space of homes in the United Kingdom and about double the average home size in France.

Can't do without space for your kids or guests to make noise? Consider converting part of your shed into an independent play room/ guest room. It only needs heating and cooling when it's being used and it'll force you to do something about the stuff in the back of the shed that's never used.

ELECTRIC HOME

A lightbulb turns on, the world's sustainable future becomes clear... everything is electric. Our home heating, home cooling, cars, trucks and even planes are all powered by electricity, generated from renewable power stations. Converting to electricity is essential for a carbon positive world.

If you ask energy experts, this future is a real possibility in just 20 short years. Excited? Here's what we can do, to get on board with a sustainable future that's electric.

1. Get solar panels

There's plenty of businesses offering to install solar panels on your roof. Some even offer to install the panels with no upfront charges, letting the panels pay for themselves over a multi-year contract. Installing solar panels makes economic and environmental sense. The investment pay off period can be 5–8 years, depending on your roof orientation and shading.

2. Detox from gas

There's a network of underground gas pipes in almost every metropolitan area. Being supplied with gas was once as essential as being supplied with water. Not anymore. New housing developments are being built for electricity in some parts of the world. There's no gas infrastructure because everyone's energy needs will be met with electricity. Think electric heating and cooling, electric hot water and electric stove tops.

In addition to the greenhouse gas emissions from using gas, there are also environmental impacts from extracting gas, including pollution of waterways and fugitive greenhouse gas emissions.

3. Electric in the garden

The humble lawn mower is essential equipment for homes with a grassy lawn and a white picket fence. Saturday mornings at home in summer are filled with the hum, or blaring noise, coming from lawn mower engines. That noise pollution is no longer needed… enter the electric lawn mower or a front yard that's landscaped with native shrubs. There's a heap of electrical equipment on the market for use in the gardening, including lawn mowers, whipper snippers and leaf blowers.

GREEN TRANSPORT

Zip, zoom, zip. Just squeezing in a trip to the shops. Just picking the kids up from swimming lessons. Just heading to the gym after work. Let's go!

We're all busy going somewhere. We all need to get around. Sometimes it feels easier to use the car.

Our reliance on cars comes at a cost. Greenhouse gasses from private vehicles are a significant and rising portion of the world's greenhouse gas emissions. Cars are also responsible for large amounts of inner-city land use. Think parking lots and roads. For people who live in the city, cars are noisy and create localised air pollution.

The good news is that there's a way around cars. Cycling, walking and public transport are good for your health, and the health of our planet.

Cycling

Bicycling is the most energy efficient form of transport. On average, it uses less energy over distance than walking. It uses a renewable fuel — your energy. There are also no greenhouse emissions or petrol costs. Ask your local cycle club about off road cycling routes in your local area. That way you can enjoy the scenery, get fit and get where you need to be going!

If you're new to cycling, getting started is as easy as 1, 2, 3!

Step 1: Find a bike that you can LOVE.
Step 2: Know your cycling wardrobe. Do you like to ride in lycra? Others like to ride in regular clothes, just not tight-fitting suits or dresses. Consider a brightly coloured cycling jacket or vest.
Step 3: Gear-up your bike. Make sure you have lights that are easy to charge, a good helmet and a lock.

Walking and public transport

Walking is also a great way to get around. If you live too far from town to walk both ways, consider walking into town and catching the bus home. The bus can do the heavy lifting, while you enjoy the walk.

Better with a car

There's lots we can do to reduce the impact of car journeys. Does your household own two cars? Sell one of them and reap the benefits of alternative transport, while keeping the flexibility that owning a car brings. If you live in a city, consider swapping a car for a car share scheme.

Get more out of each car journey. If you drive to work, can you carpool? Maybe your office will offer preferential parking to encourage you. If you drop kids at school, can you drop them part-way and let them walk the last part of the journey? Each time you get in the car, plan to get as much done as you can. One big trip is more efficient than four separate trips.

If you're choosing a new car, take a look at electric and hybrid options with a renewable power source. It's been just over 20 years since the first hybrid electric vehicles hit the market around the world. Initially, they were smaller model cars with a 100km driving range. Today, there's every type of car available with an electric engine… even a Tradesperson's ute and fancy sports cars.

Electric vehicles are here to stay. There are already charging stations around the world, with motor vehicle organisations and governments committing to expanding the current network even further. France, Germany, UK, Norway, Ireland, Netherlands, Israel, India and China are all banning petrol and diesel vehicles from 2025–2040. It's not a matter of if, but *when* your next car will be electric.

SHOPPING: BUY NOTHING NEW

Our lives are full of stuff. Some of it we use all the time, some we use just occasionally and some, well frankly, is just junk that shouldn't have been bought in the first place.

We're not alone in our lives with too much stuff. 'Australians spend more than $10 billion each year on goods we never use, clothes we never wear, and food that we don't eat,' says Clive Hamilton, Professor of Public Ethics at the Centre for Applied Philosophy. Australians send nearly 20 million tonnes of waste to landfill each year.

So how can we turn the tide? Join the buy nothing new revolution! Take up the challenge to buy nothing new for a day, a week, a month or even a year!

Australian woman Sash Milne and her two-year-old daughter took up the challenge. Sash succeeded and says that she is now more mindful and more connected to her community. She believes that taking on a buy nothing challenge can teach us to 'be present, to be mindful of the way that we behave, the words that we use… and the things that we buy.'

Milne isn't the only person who's embarked on a buy nothing new challenge. There's a quite a few people who've taken the challenge and everyone has their own tips for success. Sustainable fashion guru, Jennifer Nini says to 'avoid shopping malls'.

American Molly Knox's best tip is to avoid buying anything on credit. If you can't afford it with cash, don't buy it. Milne suggests that you head to the local library for 'new' books or children's toy libraries for toys. There are exceptions to the buy nothing new theme and most agree that food and medicines are excluded.

To get stuff that isn't new, there are a few options. There's the traditional option of thrift stores which stock mostly clothing, furniture and other household goods. There's also council-run reuse centres. These often have large household items, furniture and a large range of outdoor and gardening equipment.

Things have been taken up a notch with local 'buy nothing' Facebook groups being established around the world. These offer people a way to give and receive, share, lend and express gratitude through local networks. If you need something, ask the group – a piano, tennis racquet or kids soccer shoes. If you've got something to share, offer it up — a glut of zucchini's, pots and pans, furniture.

So set yourself a goal — buy nothing new for a day, a week, a month or a year. You'll reap the rewards by feeling closer to your community and have more money in the bank for things that you really do need.

4. Waste less stuff

Every piece of stuff that we buy new was created from our planet. We extract minerals from the ground, grow crops for food and fibre, manufacture and transport — all using energy and water. When we create less stuff, there's a global energy and water saving. Try a buy nothing new challenge... for a week, a month or a year (excluding food and essential medicines) and see how this changes the way you think about new things. You can still source the things that you need to live your life, just buy second hand using an online platform like gumtree or Facebook Marketplace, join the buy nothing Facebook network to give and receive free stuff and scour local thrift stores. Live simply and be inspired by Marie Kondo's clutter free life.

SHOPPING: SMART MONEY

Money makes the world go round. Where would we all be without access to a bank account, a home loan, insurance or retirement pension? The choices we make with our money can influence far more than just our personal bank balance and convenience — they're fundamental to our planet's health. If we're not careful, every time we put money aside for retirement, we could be investing in fossil fuels, deforestation, human rights abuses and gambling. If we care about more than just our financial return on investment, it doesn't have to be that way.

Here's how to be smart with your money and get it to create the future planet that's sustainable and regenerated with just a few, simple choices.

Retirement pension

According to UK-based charity, Make My Money Matter, switching your pension fund to one that is environmentally and socially responsible is 21 times more impactful than switching to renewable electricity, switching from air to rail travel and becoming vegetarian. Pensions are the largest single asset pool in the global market, contributinga massive USD 56 billion to global investments in 2020. A few small decisions from many individual humans can make a big, lasting difference.

Get to know your retirement pension arrangements. Most developed countries have a compulsory retirement pension scheme, where employers contribute a percentage of your fortnightly income to a fund. Employers usually have a default fund but you can also choose. Do your research and choose a fund with environmental and social credibility.

Banking and insurance

As consumers we have the power to set our expectations of the future by voting with our money. We can choose our bank or insurance provider, based on both customer service and climate impact.

According to the International Energy Agency, the road to global net zero emissions by 2050 is narrow. Along that pathway there must be no new oil and gas fields approved for development and no new coal mines or coal mine extensions. Unfortunately, governments are failing to adequately regulate and the surest way along this narrow pathway is to follow the money... by setting our expectations as consumers and making it clear that we won't be customers of institutions without an environmental or social conscience.

Want to make sure you haven't been green washed? Companies that are certified B Corporations have demonstrated high social and environmental performance. There are banks and insurance companies around the world already signed up.

Ethical investment

Ok, so not everyone has enough money to be able to invest... If you do, make that money work to create your vision for the future. Consider investing with a fund or company that has a positive environmental or social impact. Generally, the world of ethical investments simply excludes particular industries, like fossil fuels, gambling and weapons production. You can take it one step further and invest with companies that take a positive-vetted approach, only investing in climate positive, sustainable and regenerative projects, like Australian Ethical Investment.

Household level consumers can drive standards across the finance sector, from banking and insurance to pensions and investments. Let your money work to create a sustainable future for us all.

SHOPPING: ETHICAL SPENDING

Marie Kondo's method for tidying offers a beautiful approach to shopping. Before you purchase, can you visualise how that item will create joy in the life that you imagine for yourself? Extend this one step further and ask if that item will create joy for our planet? With household consumption responsible indirectly for 60% of the world's greenhouse gas emissions, reducing consumption and making ethical choices can (almost) solve the climate crisis on its own.

Here's how you can get on board with ethical shopping.

1. Less is more

Before you buy... do you really need that item? Does that item really spark joy? Is it simply an impulse purchase or something that you've thought about needing for a long time. Can you simply borrow the item from a friend or live simply with what you currently have. Can you rent a piece of evening wear, rather than buy something new for a one-off occasion?

2. Love local reuse

Get on board with the reuse revolution and extend demand for locally reused products. Search out your local pre-loved clothing boutiques, crafty furniture upcycling businesses and thrift stores. Go online for pre-loved exchanges that are free, using the Buy Nothing network, or involve a financial transaction, like Gumtree or Facebook Marketplace.

3. Search out accredited ethical brands and entrepreneurs

The good shopping guide is a UK-based book and online guide to ethical shopping. Brands are rated across categories that includes their impact on the environment, animal welfare and society. There are recommended brands for everything from kitchen appliances to insurance.

To be confident that the whole company is focused on systemic social and environmental sustainability, shop from a certified B Corporation. To be accredited, companies must meet the highest social and environmental standards, as well making legal commitments to stakeholders and be transparent in their reporting. The growing B Corporation movement includes more than 4,000 accredited corporations in 70+ countries. Global brands like Patagonia, Ben & Jerry's and the Body Shop are on board, along with a myriad of local entrepreneurs.

In the shop already? The Fairtrade label leads in connection between farmers in developing countries and consumers in the developed world. It's co-owned by 1.8 million farmers and offers products that include coffee, textiles and sports balls. Look out for the Fairtrade label.

4. Treat yourself with services and experiences, not stuff

A new pair of shoes or a dress can make you feel beautiful on the outside. How about a massage that gives an inner glow? Generally, local services and experiences have a much smaller impact on the planet than internationally produced stuff. Consider a massage, a visit to your local art gallery, a walk in the Botanic Gardens or a trip to the movies instead of shopping.

Before you buy anything ask a simple question, will this item spark joy for me and for our beautiful planet?

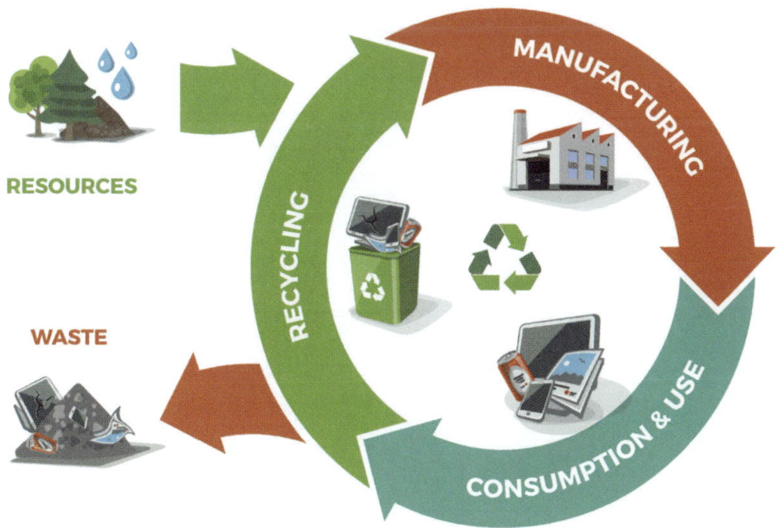

'Those shoes look amazing!' Instantly you're feeling good about your recent purchase. These shoes are special to you. They have blue leather with fancy stitching and a rubber sole. They were made in a factory on the Yangtze River delta, in China. The price you paid in store doesn't include an offset of carbon emissions or repair damage due to the die and leather processing in China. The shoes are a wardrobe staple and you wear them regularly for a year, a little less the next year. Then, it rests at the back of your wardrobe for a further year until they're a little mouldy and are finally tossed in the bin... destined to take 40 years to break down in landfill.

FOOD: SMALL FOOTPRINT EATING

There's a world of diets that make us healthier, slimmer or both. How about a diet for the planet? Food production is responsible for one quarter of global greenhouse gas emissions. The food we eat can really make a difference. There are two main parts of our diet that impact the planet: protein intake and food production. While both have an impact, it's our choice of protein that can make the biggest difference.

Sustainable protein

Tofu, nuts, pulses and avocados are all great sources of protein. According to scientists, these also have a low greenhouse gas impact on the planet, especially compared to meat. Easy, the low impact diet for our planet is to be vegan or vegetarian! Not so easy. While there are differences between countries, only a small portion of people around the world are vegetarian and even less are vegan.

So, let's explore a middle way. Here are four ideas for eating with a smaller protein footprint on the planet.

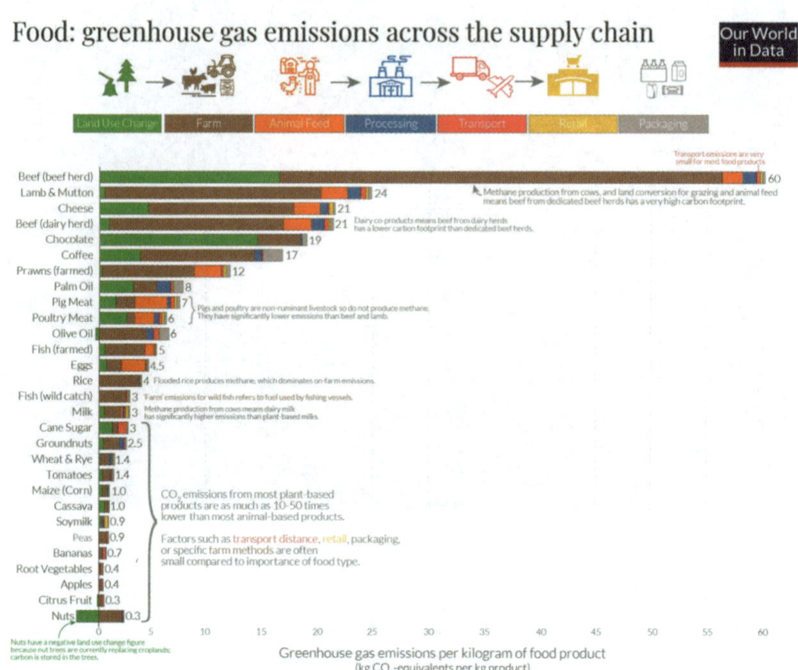

More vegetarian meals

The meat-free Monday movement tries to get people thinking about eating vegetarian once a week. You could try for three or four meat-free evening meals each week. Have you thought about the Indonesian dish Gado Gado with a solid protein offering of delicious peanut sauce, boiled eggs and tofu as well as lots of veggies and potatoes? As the springtime blossoms, try minestrone soup, a complete meal that's packed with spring veggies, chickpeas and potatoes.

Less cheese

Less cheese? It doesn't seem right. Cheese is a yummy source of protein and it's a vegetarian option. Unfortunately, because of the large amount of milk used to make cheese, and the methane produced by milking cows, cheese has a higher greenhouse impact than the meat from chicken or pork.

Think of wild fish

Fish can be the most sustainable animal protein, when it's wild and from a sustainably managed fishery. It also contains a heap of great proteins from a nutritional perspective. Try your local farmer's market or choose fish from a country that manages sustainably. Look for certified sustainable fish, including the Marine Stewardship Council.

Ethical meat

If you're going to eat a little meat then think about choosing local and ethically raised options. Some producers have already offset emissions on their farms, look out for carbon neutral meat products. Chicken has a lower impact on the planet than lamb or beef. So, if you can, substitute chicken for red meat. If you're after the best chicken, try free range or even better, pastured chicken with less chickens per area than free range chicken. Pastured chicken has a totally different flavour to free range, you've got to try it to know.

Food Production

Eat local

A typical family's healthy food basket from the supermarket with fruit, vegetables, bread, eggs, meat, cereal and some treats will have travelled halfway around the world before they end up on our plates. Almost all of this travel was powered by fossil fuels and contributing to global warming. The way your food travels matters. Highly perishable products like lobster or raspberries that travel by air have a much bigger carbon footprint than dry goods that travel by road, rail or ship. Food transport contributes just over 3 billion tonnes of greenhouse gas emission each year.

Eat seasonal

Get to know your local farmer's market or farmer's 'outlet' store. Sourcing directly from farmers is a great way to improve freshness, quality and creativity in your fresh food. It will automatically force you to eat with the seasons, reducing the need for hot houses and cold storage that are often powered by fossil fuels.

Reduce processed food

Processed food usually includes more than one ingredient. So, all the ingredients have to be transported from where they were grown to the processing plant. Then they're transported again to your local shops. They also come packaged. This all makes a big difference to the food miles in the finished product.

Consider making your own muesli at home or swapping to a simple breakfast of just oat porridge. If you've got school lunches to make, bake your own muffins and store them in the freezer. Get creative and make your own pickles, jams and other items from the seasonal recipes and home-made kitchen section.

> **Embrace the planet diet and eat your way to a healthier you and a healthier planet.**

Waste less

Loving our food by wasting less and buying what we need can also reduce our footprint. A typical household in Australia throws out 20% of the food that comes into the kitchen, either through spoiling or not eating cooked meals. There's an opportunity to reduce the food footprint by 20% just by planning and eating with intention.

For more tips and inspiration on wasting less food, look at the *Love food, live without waste* section on page 65.

FIVE WAYS YOU CAN ADAPT TO OUR CHANGING CLIMATE

We're officially living in a warmer climate. According to the UN, our planet has heated by more than 1 degree, since the industrial revolution. Overall, most places will simply get hotter. In addition, climate modelling expects wet regions to get wetter and dry regions to get dryer. It also expects wet times of the year to be wetter and dry times of the year to be drier. It's time to re-imagine the way that we live in our homes and with our communities as the fabric of our planet is changing.

1. Insulate against temperature extremes

To create an oasis at home, no matter what the temperature, you can insulate the ceiling, walls and floor. There's almost always room for more insulation. Don't stop until you've run out of space to put your insulation materials. If you're insulating, you'll also want to seal up the cracks that leak hot or cold air.

2. Supply your own water for the garden

Keen to create an oasis of veggies, flowers and space for native animals to enjoy? Climate change will make the extremes bigger with dry areas getting drier and for longer periods. Luckily, we have a few options. A rainwater tank is perfect for people who want good quality water that can be directly applied to plants. Capturing your grey water — the water that you've used to shower, bathe and wash clothes is a perfect way to make use of water that would have been treated and sent out in a local waterway. Grey water is not for your veggie patch, but it's perfect for plants that you don't eat. Capturing grey water is also pretty cheap and easy — costing as little as $30 for a hose that connects to the end of your washing machine.

3. Be an emergency services volunteer

Rural fire fighters and state emergency services are staffed by the community. There are perks — the training is pretty cool and you'll meet other like-minded people. You can also be confident that your volunteering time will make a difference to life and death situations. We're going to need many, many more volunteers as our climate continues to change and natural disasters become more frequent.

4. Love our wildlife

Better understanding populations of wildlife is not just the domain of scientists. Citizens are increasingly being welcomed to the team with programs like FrogWatch, Natures Notebook and Climate Watch. You can be part of the solution to climate change adaptation for wildlife, without needing to work in the field full-time.

5. Be part of the movement for climate change action

Connect with organisations likeone million women, 350.org and student strike for climate action. One million women was founded by Natalie Isaacs, who started out as a cosmetics manufacturer. She had an epiphany and turned her life upside down. The student strike for climate action was founded by Greta Thunberg who wanted to send a message to decision makers that young people care about climate action. Add your voice and your money, so that these movements can be stronger.

We can create an oasis in the changing climate, it just takes a little creativity and the desire to change.

COMMUNITY MICRO FORESTS

Birds chirp and children play in the dappled shade. A patch of grass, weeds and dirt between some houses has been transformed into an urban community micro forest. It brings hope at a time when hope for a better planet is in short supply.

The community micro forest movement is fast building momentum. It's a beautiful way to connect communities, enhance biodiversity and cool the landscape. Urban micro forests are emerging in school grounds, city centres and in the suburbs.

Here's how to get your own micro forest project started.

Step 1. Find a team of forest-passionate people

Work with existing community organisations, like the parents and friends of your local school, community councils or a local environment group. There might even be a micro forest group nearby. If the right group doesn't exist, don't despair — create one! A core of 3–5 passionate people is enough to get started.

Step 2. Consult with your community

Identify options for the location of your urban forest and consult with people who live and use the space around those locations. Schools and universities often have a simpler decision-making process for unused land, whereas cities and suburbs typically have more complex requirements. The best consultations are open and welcoming to a broad range of people and include targeted conversations with decision makers and community influencers.

Step 3: Raise the funds

Crowd funding is a great way to let your community know about the project, as well as engaging them in kick-starting it. Drop a postcard into the letterboxes of people in the local area, inviting them to come along to a planning session or make a small donation. Local businesses may wish to sponsor your project. Consider targeting businesses that could provide landscaping services or look for local philanthropic organisations. Community grants from government are another option.

Step 4. Technical design

Starting with the ideas that emerged from your community consultation, work with experts to create a landscape plan. Experts can help with species selection and may add design ideas. Consider whether your budget extends to earthworks for changing the landscape, or just to dig holes for plants. Do you want a dry creek bed as a feature? Do you need to improve the soil? Does there need to be space for people to gather and enjoy the forest?

Step 5. Approvals

Now that you have a design, obtain the right approvals. For a school or university campus this is likely to be simple. For unused urban land, the process can be lengthy and may require tenacity.

Step 6. Get planting!

Advertise widely to create an event where everyone feels welcome. Consider pre-digging holes for your plants, to make things easier for volunteers. If your budget extends, include food and drink for volunteers so that everyone can meet each other after the planting is completed. If you don't have enough volunteers to plant, consider whether other ready-made community organisations, like the scouts, might like to volunteer for you.

Step 7. Maintain and celebrate

Watering and weeding are constant. Micro forest maintenance requirements depend on your climate, the time of year when you plant and the species selection. Recruit a regular maintenance team and set up a roster. Create working bees that bring your community together. If volunteer maintenance doesn't get on top of the jobs, consider a sponsorship arrangement to share maintenance with a paid gardener. Celebrate planting milestones with your community as the forest grows. If there's space, continue adding to the forest.

ZERO WASTE JOURNEY

The power to transform rubbish into new resources is ours. The technology is here, right now. We simply need to use it. We need to change the way we think about 'stuff', the objects that we use for a moment, a day, a month or a year, only to be discarded as rubbish.

Imagine if all of your rubbish was stockpiled around your home, instead of being taken to landfill by the garbage truck each week. If you (almost) fill a 140L bin, each week, then every six months you'd have a pile of rubbish that looked like a mid-sized car. After a year, that's two cars. After two years, that's 4 cars filled with rubbish. A smelly, disgusting and toxic mess.

Actually, that pile of rubbish isn't just rubbish. It could be considered a resource. If it's separated the right way then much of our rubbish can be reused or recycled. There's gold and other rare metals in our e-waste. Our food scraps can transform into fertiliser for our own veggie patch or local farms. Soft plastics can be reprocessed to make new plastic or crude oil.

Let's consider the useful life of some common household objects:

Plastic straw	10 minutes
Cling wrap	1-2 days
Toothbrush	2 months
Smart phone	2-3 years
Computer	3-5 years
Business shirt	35-50 washes, approx. 2 years
Particle board furniture	5-10 years
Ball point pen	900 metres (in comparison a pencil writes for about 5km)

Consider, for a moment, the story of these manufactured objects. Usually, they've started life as raw materials like metals in the earth, oil, trees or a food crop. This raw material is extracted, transported, refined and then combined with other materials to be manufactured into the stuff that surrounds you. After stuff has been used, it's then discarded to landfill to see out the next 500 or more years. Or… if it's a lucky object, it might be recycled into new stuff. Most of the objects that we manufacture spend longer on their journey of raw material extraction, manufacturing and transport than making a useful contribution to your home.

All is not lost. With current technology, it is possible to recycle more than 85% of the waste that's produced globally. There's also plenty more that we can do at home to consume less and reuse more around the home.

Time to start on a zero waste journey. Every journey, no matter how long, starts with a single step. It's a great way to get started on the towards a zero waste aspiration. Feel good about each step that you make towards a zero waste life.

Step 1. Current waste situation

Understanding the waste that you currently produce is the first step towards reducing your waste footprint. So, shortly before your weekly garbage collection, take a look at the contents inside your bin. If you can handle the smell, tip it out and sort your rubbish into the below categories and give each a percentage value, with the total adding up to 100%.

Food and green waste	
Paper and cardboard	
Plastic	
Glass	
Metal	
Rubber and leather	
Wood	
Other (e-waste, nappies, etc)	
TOTAL	**100%**

Ok, if it's too smelly or disgusting you could simply rummage through your bin and try to estimate the percentage value of its contents. Don't forget to factor in the things that don't get thrown out each week, like E-waste.

A typical waste stream in Australia, the US or UK, before recyclables are removed, looks a bit like this:

Food and green waste	32%
Paper and cardboard	25%
Plastic	13%
Glass	5%
Metal	6%
Rubber and leather	4%
Wood	4%
Other (e-waste, nappies, etc)	11%
TOTAL	**100%**

So, if you're a typical household, most of the rubbish bin that you produce can be recycled. Here's how.

Step 2. Recycle like a pro

Ok, so we're all good at filling up our recycling bin. But, it's not just about filling up that bin. Being good at recycling is about making sure there's nothing in your rubbish bin that should be in the recycling bin. In many houses, some rubbish in their landfill bin is simply stuff that wasn't correctly sorted.

Unsure? Your local council will have advice on their website about the items that can and can't be recycled. One of the most common recycling errors is for people to place recycling inside a soft plastic bag. Unfortunately, in this case, the whole lot will go directly to landfill.

Mastered that? You're ready to take your recycling to pro standard. Here's how:

- Remove soft plastic from your rubbish bin and take it to a recycling drop-off location. There are hundreds of places that accept soft plastics,

Separate soft plastics from waste that can't be recycled at home.

many are co-located with retail outlets and supermarkets. Recycling soft plastic makes an enormous difference to the amount of rubbish in your bin. For many households, soft plastic is 20% of the content of their bins. Any soft plastic that passes the 'scrunch' test can be recycled, including bread bags, cereal packets, biscuit packets and broken 'green' bags. The list of items that can be recycled is simply enormous. Check out the online guidance from your local drop-off location, it'll make a huge difference to the size of your waste to landfill.

Set aside a separate space for collecting soft plastics in your home and make soft plastic recycling a new habit.

- Use the container deposit scheme in preference to your curb side recycling. Why? The quality of materials recovered from the container deposit scheme is better than the quality recovered from curb side recycling. Better quality means that the new material can be used in more stuff. So, instead of simply being able to make hard plastic items from the raw material, manufacturers can use material from the container deposit scheme to create almost any new plastic object. It also earns you money. If you don't want to drop off the objects yourself, check to see if any local community groups offer a collection service.

- Get more out of curb side recycling by grouping 'like' materials. Because of the way a materials recovery facility sorts our recycling, small pieces of paper, plastic and metal are difficult to recycle. No problems. Simply group small pieces of material with larger ones.
  - Put your paper receipts and other scraps into a used envelope, then close it up before you put it into your recycling bin.
  - Small pieces of plastic can go inside a larger plastic container with the lid on... think take-away forks inside a take-away container, or small plastic lids inside a plastic yoghurt container.
  - For metal to be recycled, it needs to be about the size of a golf ball to be picked up by the magnets that sort it out. So, save up small pieces of aluminium and wrap them in some used aluminium foil to make a ball. Small pieces of steel, like many bottle tops, can be placed inside a steel can, and the can folded over onto its self with a strong squeeze of your hand. Not sure if a metal is aluminium or steel? Steel is attracted to magnets and aluminium is not.
- Know the enemy of recycling: composite materials. When an item is made of two different materials that are stuck together (usually with a strong glue) then it can't be recycled. Best to avoid buying these in the first place. Good examples of composite materials are chip board (wood and glue), medicine packaging (foil and plastic) and children's toys (plastic and electrics).
- Take used products and packaging back to the place where you purchased them if the store offers this service. Clothing giant H&M accepts clothes in-store for recycling, beauty product containers can be returned to David Jones and some home electronics are accepted for recycling by the store that sold them. Ask before you make that purchase.
- Get your community recycling with unusual recycling boxes. Commercial companies post out a box that you fill with unusual items including coffee capsules, office materials and disposable gloves. That commercial company then dismantles the items and recycles as much as possible, for a fee. Mostly, these recycling boxes are aimed at schools or businesses as the collection point.
- Recycle your E-waste. This is the fastest growing component of household waste. It also contains precious metals, like gold and platinum,

that are not renewable. E-waste includes all electrical appliance waste, whether the item is obsolete or unrepairable. Local councils and most national governments have guidelines for E-waste.

- Search out your local recycling options. Councils often have specific collections or drop-off points for bulky waste, batteries or chemicals. Community groups sometimes recycle boutique items. Lids4kids Australia rescues plastic lids from milk, juice and other bottles from landfill and provides them to schools, community groups and small business to turn into meaningful recycled products.

Step 3. Remove organic waste from your household rubbish bin

Some councils offer an organic waste collection scheme that includes both food and garden waste. If this is you… woo hoo, easy to remove organic waste from your household rubbish bin!

If you've just got garden waste collection or none at all then investigate these options for recycling organic waste. There are options to recycle organic waste that suit every lifestyle. Apartment dwellers can use bokashi, people in townhouses might choose the worm farm or a compost tumbler and those with big backyards can have chickens.

Here's an introduction to each organic waste recycling option, to help you decide which one is best for your life.

Bokashi bucket

Developed in Korea, food waste is placed in a kitchen bucket and broken down with the help of a 'bokashi' enzyme. The Bokashi bucket suits people living in a small space, like apartments. You can add almost all food waste to the bucket, except large bones and liquids like soup. Every time you add food scraps, add the bokashi enzyme to the bucket to facilitate your food waste breaking down. The bucket produces liquid fertiliser — perfect for balcony plants — and excellent quality soil.

Worm farm

When working well, having a worm farm is straight forward. Worms eat most fruit and vegetable scraps, tissue paper and old natural fibre rags. Leave out meat, dairy, citrus peel, onions and garlic. Worms don't have teeth, so they like their food to be chopped up or very soft — easy to do when you're in the habit. Suitable for anyone with an outside space, including a small balcony.

There are a few tricks to keeping a worm farm humming along. Location is important... and this isn't about smell. A functioning worm farm doesn't smell of anything, just soil. Worms need a relatively consistent temperature and suit a courtyard space, up against the wall of your home. They freeze in sub-zero temperatures and they overheat in the direct summer sun.

Compost

Composting is versatile — you can add anything that once lived into a compost that's working well. Yes, that includes meat, dairy and chunks of large vegetable. In addition to food waste, composts like to be 'fed' with other 'brown' or dry materials, like sticks, leaves or newspaper. The rough ratio of green waste to brown materials is 2:1. Most people layer their compost so that the brown layer adds air, an essential part of the compost's success. To have compost at home, you'll need a backyard. Usually, people have two working composts and between them you might need a space about 5 metres square.

There are three main types of compost systems:
Compost tumbler. These are sealed from rats and mice and suit smaller spaces, including courtyards. It's a pretty forgiving method. Simply turn the tumbler to add air into your compost mix.

Bin or box compost, connected to the ground. Because this system connects with the soil, earthworms and other soil microbes that break down waste can easily colonise your compost. If you need to add air, simply use a large stick to make three or four vertical 'holes'.

Community garden compost. These are large compost heaps, tended by community garden members. If there's a local community garden, why not ask if you can contribute your food so that there's extra compost for their plants? Ask your local council for advice on the nearest community garden compost.

Chickens

If you've got a large yard, consider chickens. They will eat all of your kitchen food scraps in (almost) any size or shape. They add rustic charm to the backyard, with their clucking and prancing around. Chicken manure is excellent garden food — just turn it through your soil and let it rest for a week or two before adding your veggies. Chickens also produce eggs — tasting delicious and helping the family budget.

Chickens take a bit of looking after. They need a predator proof shelter (something you can shut at night) with a roost and space to lay eggs, yard space to dig and scratch around, water and grain to supplement their food scraps diet.

Step 4. Taking it all the way to zero waste

Moving all the way to zero waste is about changing our lifestyle. It's about avoiding waste creation, through buying less stuff or different stuff. It's about reusing and re-purposing around your home. It's creative, because this part of the zero waste journey needs to be tailored to you.

A week of waste from Mia's family of five.

How will you know where to start? If you've taken soft plastics and organic waste out of your rubbish bin, then there won't be much left over. It's easy to see what's there. You might have a broken plate, an 'oxygen reducing' sachet, broken shoes, broken plastic toys or old clothes. The things that you have left over depend on the things that you've chosen to purchase and use.

Stick with these three buying principles and they'll help you tailor a journey that takes you all the way to zero waste.

1. Buy less stuff

Sounds simple, but avoiding or delaying the purchase of stuff can make an enormous contribution to zero waste. Big retail businesses know that most of what they sell isn't needed. That's why they go to great lengths to put things you haven't thought that you might 'need' in front of you and make it easy to purchase. Most big businesses offer sales online, in-store and on credit. Every time you make a purchase, ask whether or not you really need it. Do you really, really need it?

Delay your purchase any way that works for you. Delaying helps you to decide whether or not you really need that item. Perhaps you can delay by only purchasing when you have the cash, not on credit. You might like to budget for the month, or year... and stick to it.

2. Buy nothing new

New stuff comes packaged. Beautiful, shiny... and wasteful. Packaging is a huge contributor to waste, with around 40% of global plastic production dedicated to packaging. Simply buying second (or third) hand avoids the creation of packaging waste. Be inspired by the buy nothing new stories on page 33.

3. Buy things that last, can be repurposed, repaired or recycled

Most people buy for convenience. Big retail businesses rely on our convenient, throw away approach to stuff. We buy plastic water bottles... because we need a drink right now. Plastic, shiny toys are gifted... because they'll be enjoyed for a few weeks. We buy cheap shoes... because they'll look good this summer.

There are waste-free alternatives for most household items. Compostable bamboo items are springing up everywhere, as toothbrushes, take away food containers and bowls. Avoid polystyrene packaging as it isn't recyclable. It's used in meat trays and as packaging. There's a movement towards recyclable meat trays — just ask where you purchase your meat.

Choose repairing over buying a new replacement. Your local council or environment centre will be able to tell you if there's a repair café service. Most repair café's will take any household items, including furniture, electronics and textiles, repairing them for a small fee.

Here's where it's up to you to be creative. Take it one step at a time. Choose your first target, make the change, then move to another target. Go easy on yourself and celebrate the small things, every step of the way.

4. Shop (almost) without packaging

Buy your food ~~naked~~... in bulk. Remove packaging waste from your life. Bulk food stores often stock other waste free living items, like bamboo toothbrushes or compostable cotton buds.

Buy food directly from the famer. This removes the packaging that comes with double and triple handling by buying from the source.

Use your own containers for take-away food and at the deli, so you don't need that single use plastic. Like remembering your reusable shopping bag, the challenge is remembering to bring your container.

Buy some food in bulk and store them in re-purposed jars.

FRESH IDEAS FOR REUSING AROUND THE HOME

Creativity is the key to successfully reusing items around the home. The sky is the limit. We've just got to get in and try new ways of doing things. If we reuse more, we can create less waste, buy less stuff and live gently on our planet.

1. Toothbrushes

When the bristles are all turned out, it's time for a career change. Time to switch from tooth brushing to kitchen scrubbing. Toothbrushes are great at getting into small spaces for cleaning specialty kitchen items. They're also great for cleaning around the house, for getting into small and hard-to-reach places.

2. From bath towel to hand cloth

When one of your bath towels gets tatty, it's not time to throw it in the bin. It's time to get out the sewing machine and scissors. Create a hand towel or two from the 'good' fabric that's still contained in an old bath towel. When hand towels get tatty, time to get the sewing machine and scissors out again. Hand towels become kitchen rags. The off-cuts from 100% cotton towels can be put into the compost or worm farm, as can the kitchen rags when they finally become tatty too. Now, that's really using a towel to the very end.

3. Every kind of plastic bag

Zip lock bags, bread bags, shopping bags… they all go into a bespoke calico bag holder. Reuse ideas for plastic bags includes packing whole fruit and snacks for outings, taking dirty clothes or shoes home from sports training and for lining the rubbish bin.

4. Packaging

Bubble wrap and parcel packages are the by-products of online shopping. Fortunately, both can be reused. Either reuse bubble wrap at home when sending your own parcels, or ask your local post office if they'd appreciate some extra. Parcel packages can be reused, simple tape a small label over the top of the used address spaces and you're parcel is ready to go!

5. Children's craft

Where to start? There are so many options here! Here are three favourites.

1. Box construction

Take some cardboard packages, clean plastic bottles, add some sticky tape and you have perfect ingredients for box construction. This craft brings joy to many small children. Many a family has suffered the box construction invasion.

2. Bottle top art

The ingredients are coloured bottle tops and glue. Stick the bottle tops onto paper to make a design. Bottle tops also make great 'necklaces' if a hole is drilled/ punctured in the middle and they are threaded on a string.

3. Scrap booking

Just old magazines, scissors, glue and a scrap book are needed to create works of art. Great for developing children's coordination and learning to use scissors.

6. Ice cream bucket

Family sized ice cream comes in large plastic tubs. Perfect for feeding a crowd, and when it's finished the plastic tubs make a great storage bucket. Use the plastic tubs for food scraps in the kitchen, ready for the chickens or the worms. If there's extra, kids like to use plastic tubs to store small toys or 'treasures' from nature walks.

7. Electrical cord labels

Some people are ultra-organised around their homes. There are labels in the pantry, the shed and for each electrical cord. If organising is your thing, consider reusing bread ties to label your electrical cords. Bread tags are those tiny pieces of hard plastic that seal a plastic bread bag and keep in the freshness. The 'hole' in the tag is the perfect size to fit snugly around most electrical cords.

8. Glass jars

Short of water glasses? Look no further than the humble jam or pickle jar. Save them up and you'll have a matching set, or enjoy the hipster-style variation in size.

Jam jars make fantastic water glasses.

9. Wine bottes

Remember a fabulous bottle of wine by turning it into a decorative interior piece. Place a candle in the bottle's neck as a rustic, romantic table piece. Fill the wine bottle with water and pop some cut flowers or plants that grow in water on the windowsill. Herbs that grow in water include rosemary, mint, basil, lavender and sage.

10. Odd socks

Is there a place where all of the world's odd socks congregate? Socks are something that just seem to disappear from every home. On a rainy day, go through the odd sock pile and make sock puppets. A simple puppet needs just some heavy duty texters. Use the textas to draw some eyes and nose on the sock. If you're feeling fancy, stick on some craft materials to create hair, jewellery or even insect antennae.

PLASTIC FREE LIVING

Every year humans throw away enough plastic to circle the Earth four times. Half of this plastic is single-use and disposable. Much of it ends up in our oceans, where it's responsible for killing one million seabirds and 100,000 marine mammals every year. It's time to rethink how we use plastic. It's time to start again and build a relationship with plastic that's healthy for us and our planet.

Step 1: Refuse. Start with the big 4 — straws, water (and soft drink) bottles, shopping bags and coffee cups. Refuse other plastic items that you don't need.

Step 2: Replace. Find plastic alternatives, like using beeswax wraps instead of cling wrap. Take a look in your local bulk food store or online plastic-free alternatives store and choose products that suit you. There's everything from plastic-free toothpaste and toothbrushes to beauty products.

Step 3: Reduce. Buy (almost) nothing new. When it comes to reducing your plastic budget, for many people, even more powerful than the alternatives to plastic wrapped food is to buy nothing new. New household items are commonly wrapped in plastic, with plastic cushioning and ties that keep it safe while the items are in transit.

Make the change stick by choosing one or two products at a time to switch. Find a community of like minded people through the plastic free July campaign.

ZERO WASTE KIDS PARTY

Kids love birthday parties — especially their very own. There's joy in the planning… 'I want a Star Wars cake, fairy bread and lots of presents.' There's joy on the day… bouncing around the living room, waiting for party time, the joyful shrieking as party guests arrive. So much excitement. So many friends. So much stuff... So much waste.

Our best memories from kids parties aren't about stuff. They're about playing with our friends. The feeling of being special, just for a day. That mountain of stuff in the bin simply fades away.

Here's some inspiration for birthday party fun, without the waste.

Decorations

Think home-made, with recycled materials. Make your own bunting from crepe paper and string. Use leaves as confetti. Make your own custom birthday sign using paper and paints... also biodegradable.

Here's the hard message on decorations. We need to break up with party balloons. Yes, even latex or biodegradable party balloons. Unfortunately, they're deadly for wildlife who eat them mistaking them for food. Balloons left outside will get washed through our waterways into the ocean. Helium balloons that are released into the atmosphere explode, creating a shape like a jellyfish. Consider blowing bubbles and having colourful crepe paper streamers as an alternative to celebrating with balloons.

If you want eco decorations but don't want to do it yourself… try a specialist in sustainable party supplies. Many supply a full eco party supplies kit.

Food

Avoid food waste by making the party finger-food only, on platters that you've brought from home. Super keen? Use cloth serviettes and put them into the wash after the party.

There's a minefield out there when you consider how to provide cups, plates and cutlery. Here's some ideas to help you navigate your way to success.

- Reusable cups, plates and crockery that are washed. Don't have enough at home? Consider hiring or buying extra from your local thrift shop then returning them the following week for someone else to use.

- Paper plates and cups. Try to avoid the paper plates or cups that are lined with plastic. Paper plates can be recycled in your home compost or worm farm. If they're plastic lined, they can go into your recycling bin… but the plastic component will be separated out and is not recycled, just the fibre from the paper is.
- Bamboo plates, cups and cutlery. Perfect if you've got a commercial composting service available and just fine in a well-functioning home compost.
- Plastic cups. Check the type of plastic… 'bio plastic' can often be commercially composted but cannot be recycled with normal plastics. Most of the time, regular plastic cups can be recycled through your regular recycling service.
- Cutlery. Plastic cutlery is best avoided. Try bamboo or wooden cutlery that can recycle in a commercial composter or a well-functioning home compost.

Compost any food that's left over, either in your home compost, community compost or through a council collection service.

Gifts

Here's your biggest and most challenging target. Gifts, through their wrapping and packaging, can create the bulk of most kids' party waste. It's a challenging target because most kids expect a present from everyone who comes along to their party.

Having a sensible conversation about gifts with any child under the age of 6 might just be impossible. 'I want a super hero costume.' 'I want a LEGO set.' Often, gifts can inspire tears if they're not what was expected, with 'I've already got that' or 'I didn't want that', even when a gift is flavour of the month… last month. Having a sensible conversation with older children might just be possible.

Here are three approaches to zero waste gifts at a kids party.

- **Wishing well.** Invite your child to think of something really big that they'd like for their birthday. How about we ask all of your friends to contribute to this, rather than getting smaller, individual presents? The party invitation might read: *No presents necessary.*

There will be a 'Mountain Bike' wishing well, should your family be keen to contribute.

- **Community twist to the wishing well.** Appeal to your child's inner sense of altruism. How about we ask your friends to contribute towards a gift for you, and a charity donation? The party invitation might read: *No presents necessary. There will be a wishing well for those who wish to gift. Marcia will split money from the wishing well equally between her favourite charity and a single, large gift.*
- **No gifts.** For many children and parents this will be a bridge too far. It gives real emphasis on the party experience, rather than the 'stuff'. An invitation might read: *Please, absolutely, positively, under no circumstances should you bring a gift. We really mean it. George just wants to play with his friends. We won't be providing take-home party bags, so we'll be even.*

Take-home party bag

Putting together an affordable take home bag that's fun for kids and zero waste takes just a few small changes. Of course, these take-home gifts would all be supplied in paper bags that can be hand decorated by the party host — an extra craft activity. Here are some ideas:

- A small plant. Succulents are easy care. A small flowerpot is beautiful. Herbs might suit if you've just run a MasterChef themed party.
- Mini gardening kit. A packet of seeds and some gardening stuff make for good conversation over the coming months as the seeds grow. Micro-herbs provide a quick harvest.
- Home baked treats. Gingerbread, chocolate-chip cookies and caramel fudge will all be popular eating on the way home.
- Small book. Second hand bookstores and markets have discounted books — choose a selection for party guests.

ZERO WASTE BATHROOM

Here are six simple tips for starting a zero waste journey in your bathroom.

1. Switch to soap bars instead of liquid soap in a pump pack. You'll save oodles of packaging waste if you buy soap bars in cardboard wrapping. Most pump pack bottles are recyclable but the pump itself is not recyclable. If you love liquid soap in pump packs, switch to re-filling your bottles. Simply keep the pump packs that you already have and refill. Source your liquid soap for refilling from a bulk food store or from your local supermarket.

2. Switch to bamboo toothbrushes and put them into your home compost at the end of their tooth brushing life. You can pull out the nylon brushes, or just accept a little nylon in your compost. If you don't have a compost, you can simply bury used toothbrushes in your garden, they will break down (eventually). Bamboo toothbrushes can be sourced almost anywhere, from specialty online stores to regular supermarkets.

3. Check your local council's recycling guidelines. If they don't include specialty bathroom items, then consider a commercial recycling box. The service works when you buy a box that arrives in the mail. You then fill the box with used items and post it back to the service provider. They do the sorting and the recycling. If you're not keen to recycle at home, some dentists and specialist outlets offer oral care recycling services to their patients.

4. Say goodbye to toothpaste in a tube. Replace it with toothpaste tablets that come in a paper box. You can source these online or from your local bulk outlet. Simply chew the tablet then brush your teeth.

5. Be loyal to cotton. Source towels and bath mats that are 100% cotton. That way, at the end of their life, you can place their scraps into your worm farm and they'll be devoured and transformed into a liquid garden fertiliser.

6. Turn old towels into hand towels, washers or cleaning rags. When your towels get a hole or a stain, instead of throwing it out, chop it up. Use it as a rag. If you have a sewing machine, convert a large towel into hand towels or washers.

Zero waste is a journey. It starts with just one step. Like any journey, it will feel easy at times. Other times it will feel like you're climbing a snow-covered mountain in a snow storm. Our everyday choices can make a huge difference to the waste we create across our entire lives. There are no rules for which step you should take next on a zero waste journey. Start in your kitchen, bedroom or the bathroom. As you climb that mountain, may the snowstorm clear and the sunshine brighten your day.

LOVE FOOD, LIVE WITHOUT WASTE

Do you ever find that your cupboards are full of food but you can't find a whole meal to cook? Half-finished packets of rice paper but no noodles or sauce ingredients? Pre-made sauce packets are out of date by years. Limp pieces of broccoli, radishes and some mouldy peaches in the fridge?

According to the United Nations Environment Program, about one third of the world's food production for human consumption is wasted. In high income countries, most food waste occurs at households and supermarkets. Australians throw out 20% of the food that they buy. That's one in every five shopping bags thrown straight into the bin. Pretty wasteful for the home budget and also really wasteful for the planet.

Time to love food more and live without waste! Time to get sustainably creative with your food. Here's how to do it.

1. Plan meals and go shopping with a list

Start planning your meal quantities. Is that one or two heads of broccoli? After you've got the hang of it, you'll find it faster to translate meal plans into an ingredient list.

2. Eat from the back of the cupboard

Try to use up at least one ingredient from the back of the cupboard each week. Incorporate this into your meal planning and you'll be surprised at the diversity that this adds to your diet. Target those half-finished rice paper rolls by buying extra to have enough for a full meal.

3. Use everything in the fridge before going shopping

If you've got a regular shopping day then the day before is time to clean out the fridge. Perhaps this means a potluck veggie soup is on the menu the night before, or a potluck roast vegetable salad. Excess fruit can be poached and eaten on breakfast or frozen for later. Get ahead of that mushy, slimy back of the fridge thing.

4. Cherish leftovers like gold

The biggest crime with leftovers is to push them to the back so that they can't be seen and used. Extra soup? Frozen for a quick dinner. Leftover roast vegetables and meats are chopped and laid out on top of baby spinach as a lunch time salad.

UPCYCLING FURNITURE

What to do with old, wooden furniture? Add a lick of new paint, of course!

Step 1. Find some old, wooden furniture

Source furniture that is almost ready to be replaced, like outdoor chairs and tables covered with mould or starting to rot. If you don't have old furniture around your home, ask for some on your local 'buy nothing' Facebook group or visit your local recycling centre.

Step 2. Find some paint that inspires you

Take a look at the leftover paint from earlier projects around your house. Brighten up your world by making bold decisions with the paint's colour. For outdoor chairs think bright pink, blue or white. An indoor table might work best in the same green as a feature wall in the kitchen. Don't have spare paint at home? Consider asking friends or neighbours if they have anything in the back of their shed.

Have a think about the type of paint you're using. If the paint underneath is enamel then you'll need to choose enamel paint or give the surface a good sand back and use a primer. Worried about using indoor paint on outdoor furniture? Don't be. You might just need to add an extra coat or two, or plan to re-paint more regularly.

Step 3. Prepare the surface for painting

If you don't have a totally smooth surface, then give your furniture a light sanding as well as a wipe down. However, if you've got a smooth surface, just a wipe down will be enough.

For rotting furniture, you'll need to remove the rotting sections and pop in some wood putty to even up the surface. After it's dried, sand it down.

Step 4. Paint

Such fun! Pop your furniture down on the grass and get painting. Use at least two coats of paint to give a depth of colour and help with preserving the furniture's structure.

UPCYCLING CLOTHES

In a world where most things are mass produced to be cheaper than cheap, unique items have special value. They speak of character and celebrate everyone's unique personality.

Finding unique items can be a challenge. Unless, of course, you make your own.

Enter… five easy upcycling ideas. You don't have to be a sewing expert or have a spare week to do any of these projects. Three of them don't even need a sewing machine. Best of all, because these are upcycling projects, they're a great way to enjoy the fabric and memories of your favourite clothes, long after the original item is worn out.

1. Cloth shopping bag

Old jeans or t-shirts make great shopping bags. For the jeans, use the section between the waist and crotch. Simply cut this off above the leg line, put the 'right sides' together, pin along the crotch area to make the bottom of the bag and then sew and overlock around the bottom edge. For straps, use the legs of the jeans to create straps. For the t-shirts, use the section between the underarm pit and the waist. Simply cut off the t-shirt between the lower edge of each sleeve, turn the t-shirt inside out and pin along the cut edge. Sew and overlock along the cut edge to create the bottom of the bag. Create straps from the left over fabric or use a thick ribbon. Attach straps.

2. Beeswax wraps

You'll need a light cotton fabric as the base material for the beeswax wraps. An old sheet or summer shirt are both suitable. A small wrap is about 18cm x 18cm. The only other ingredient is beeswax (about 15g for a small wrap). To make the wraps, you'll need a kitchen grater (unless the beeswax is pre-grated), old iron and baking paper.

Firstly, cut your fabric to sandwich wrap size. Then grate your beeswax using a kitchen grater. Working carefully, place your fabric on top of a sheet of baking paper. Sprinkle the grated beeswax on top, so that the fabric is loosely covered in grated wax. Place a second sheet of baking paper on top of the grated wax. Run a medium iron over the top of the second sheet of paper to melt the beeswax into the material. Take care that the melted wax doesn't drip outside your baking paper layers and onto an ironing board or your iron.

3. Flower with button

Cut a favourite piece of cloth in a circle shape, about the size of your hand with the fingers outstretched. Hand sew, with large stitches (about 3cm each), around the outside of the circular shaped cloth to create a draw string. Instead of tying the knot, pull the draw string tight. You will now have a ball shape, with the draw string at the top. From the top, squash the draw string ball so that it is flat. This creates the flower shape. Using small stitches, sew the middle of the flower (where the draw string is) to the fabric at the back. To finish the middle of the flower, sew a fancy button over the draw string area so that the edges of the fabric are hidden. Attach a safety pin to the back, so that the flower can be pinned onto shirts or skirts. For a truly spectacular finish, consider making a few of the flowers and pinning them on to a shirt or skirt together.

4. Fancy patches on pants

This is upcycling for creative people who are happy to wear their personality. Pants usually wear at the knee… unless they've been ripped somewhere in the buttock region. Instead of throwing them out, hand or machine sew a patch over the ripped section. Make the patch a feature by using bright or floral material. While you're at it, add a decorative patch to complete the 'look'. Consider hand sewing fancy buttons or even decorative flowers to make your personality shine.

These shorts have it all, a pretty patch, sewing that draws together a rip and an intentional 'hole'.

5. Children's 10-minute skirt from old t-shirt

Take any old knit shirt. Larger shirts will produce fuller skirts. Cut the shirt between the underside of each sleave. The bottom of the t-shirt will be the hem line for the skirt and your cut line will become the waist. Overlock the cut edges of the t-shirt. Then fold and pin the overlocked edges of the fabric to create a 1¼ inch space for the elastic band around the skirt's waist line. Sew along the pinned edges, leaving an opening of about 1 inch for threading the elastic. Measure and cut some 1 inch thick elastic to the length of your child's waist. Using safety pins, thread the elastic through the skirt's waist band. Pin and sew the elastic, using a zig zag stitch. Then close up the 1 inch space in the waist band. Voilà! Was there ever a girl that didn't need a new skirt?

HOST A CLOTHES SWAP PARTY

Do you look into your wardrobe full of clothes and think there's nothing to wear today? Do you look enviously at the clothes on your friends and colleagues? Do you like to look nice without the environmental or social impact of our global fashion industry? Time to organise a clothes swap party.

A clothes swap party is a great way to empty your wardrobe of unused or wrong-sized clothes, take home some 'new' second hand clothes to your wardrobe and socialise with friends. Organise it as an afternoon with cake, champagne and friends. The ultimate clothes swap party has 15–20 people.

Here's how you can host a clothes swap and make it a success.

1. Invite friends with all different shapes and sizes. That way there'll be something for everyone.

2. Schedule the swap when there's a change in season. Your friends will already be looking for new clothes to wear and you can provide the perfect excuse for them to refresh their wardrobe. Let people know a few weeks in advance, so they have plenty of time to decide what they'll contribute to the swapping pool.

3. Ask everyone to bring clothes that are in good condition. Clothes with holes or bad smells should be left at home.

4. Add shoes and accessories to the mix. Including handbags and jewellery means that everyone can take home something, regardless of their size.

5. Eliminate clutter from the swapping area and designate places for different types of swapping items. Use a portable clothes rack for suits and work clothes, fold up T-shirts and put them on a table, drape the dresses over your lounge, shoes and accessories on the coffee table.

6. Designate at least one room for changing and have at least two full-length mirrors available for people to see how their new clothes look.

7. Put all of the clothes out before you start swapping. That way people who arrive 5 or 10 minutes late don't miss out on the 'nicest' clothes.

8. Encourage people to use the clothes swap as an opportunity to take fashion risks. Take home things that you've always wanted to wear but haven't had the courage to buy. Help your friends to choose clothes that you think will look good on them. It's a party!

9. At the end of the swap, you'll have left over fashion that's dated and hasn't come back again… yet. Put all the leftovers into large garbage bags and donate it to a charity of your choice.

10. Like the idea of swapping but not keen to organise your own event? You can swap high quality clothes online through organisations like the clothing exchange in Australia and Canada.

SUSTAINABLE GIFTS TO MAKE AT HOME

Gifts are a great way to let people know they're appreciated. They say thank you in a way that people will remember.

What better way to do this than to make my gift at home? It's an easy way to infuse love into every gift. Best of all, there are plenty of simple gifts that you can make in bulk. No more last-minute trips to the shops, just grab something from the cupboard that already comes with love!

Here are six fresh ideas for gifts that you can make at home.

1. Strawberry Jam

Scones with jam and cream and sponge cake with jam filling, delicious! Strawberry jam is the best. Here's how to make it. Ask at your local farmers market, farm gate or grocer for jam making strawberries. They are often sold by the tray. Trim and wash the strawberries then roughly chop them up and put them into a saucepan. Add an equal weight of sugar, the peel from a lemon and vegetarian pectin. Adjust the amount of pectin to the weight of your fruit and sugar. Bring this mixture to the boil for 5–10 minutes, skim off the 'froth' and bottle in sterilised jars. If it doesn't set, don't worry... it's strawberry pancake syrup!

2. Lemon butter

Vintage. Love it! Easier to make than you think... the trick is patience. Firstly, prepare ½ cup lemon juice, two teaspoons of grated lemon rind and 125g of cubed butter. Set them aside. Whisk 4 eggs with ¾ cup of castor sugar on top of a double boiler. Keep whisking until they're well combined and the mixture thickens (this is the patience bit). Then, as soon as the mixture thickens, add the lemon juice, lemon rind and butter. Whisk until it's well combined. Bottle and store in the fridge. Best eaten with strawberries and cream.

3. Caramelised balsamic vinegar

Amazing on salads or vegetables. It's strong, sweet and easy to make. Bring three cups of balsamic vinegar to the boil. Stir through two cups of brown sugar then reduce to a simmer. Simmer for 30–40 minutes until the mixture has reduced by half. Place in small, sterilised bottles.

4. Gingerbread

This is a great Christmas gift that children can make for their friends. Combine 125g of melted butter with half a cup of brown sugar and an egg. Stir through 2½ cups of plain flour, 1 tablespoon of ground ginger and 1 teaspoon of mixed spices. Kneads into a dough then let it rest for about 30 minutes. After the dough has rested, roll it out between two pieces of baking paper and cut it into fancy shapes. Decorative icing is optional.

5. Pot of herbs

Nothing says abundance better than a pot overflowing with greenery. Reuse pots from around your home, fill them with a combination of garden soil and worm casings, then plant a heap of seeds on top. Basil, coriander and parsley are all easy to grow. They also produce a nice bush. You'll need to remember to water the pots — about three times each week — so place them somewhere that you walk past regularly. Seedlings are truly a gift of love. Cultivated by hand and nurtured as they grow and flourish!

6. Microgreens

This is the fast, hipster version of a seedling gift. You'll need a shallow, wide tray… or a wider pot. Prepare a rich, well-drained soil from around your garden or source commercial potting mix. Then cover the top of the soil with seeds — I mean really cover it! Because the seeds will be harvested young, you can plant seedlings really close together. Favourite microgreens include red cabbage, rocket, basil, coriander and English spinach. Sprinkle a thin layer of soil on top of your layer of seeds. Water your seeds regularly. In late spring and summer, microgreens are ready 4–6 weeks after you sow the seeds.

ECO BABY

If you're shopping for baby gear, you'll find gadgets for almost everything. Thermometers in all shapes and sizes, a video link to your phone, games that learn with your baby and more.

What you won't find at the shops is the simple approach to living sustainably with your baby. This approach is tucked away in people's houses and in private conversations. Here's that private conversation about getting ready to live sustainably with a new baby.

Baby gear

There seems to be an expectation that parents will buy a mountain of gear including a pram, bassinet, cot, change table, new set of drawers, travel cot, baby carrier, car seat, new clothes, thermometer and more. The Baby Centre estimates that the cost of a baby's first year can be about $10,000. Is it all essential? Does it all need to be new?

Everyone has a different set of furniture and equipment that suits their lifestyle. Most people will tell you that a change table is essential. Not always. A baby's nappy can be changed on a mat at the end of your bed.

Baby bath essential? Maybe not if you choose to get into the adult sized bath with your baby, or if you want to shower with your baby from day one. Have a good think about what you really 'need' and what is essential for your lifestyle.

Once you've decided what you need, then buying it doesn't have to be a drain on the budget or a festival of shiny, new plastic. Babies rarely wear out their bassinet, cot or change table. So it's usually easy to find good quality, second hand.

Try these places for second hand gear:

1. Baby and Kids markets. These bring together parents with second hand baby furniture, clothing and toys to sell, with parents who need to buy. It's best to get there early to have your pick of the bargains.

2. Gumtree and all classifieds offer an online service for buying and selling. Both have a heap of items in the baby or family categories that you can pick-up locally, or further afield if you're prepared to travel.

Food

Breastfeeding doesn't need any special equipment and is a great way to feed your baby sustainably from the start. There are free national help lines, an online chat service and other great resources for parents in Australia, the UK, the US and Canada. You can even take classes before your baby comes, to build confidence in the real thing.

When you decide to introduce solid food, don't feel like you have to buy baby-specific pre-packaged items. It's easy to make your own food at home, just keep it low salt and full fat. There's also the baby led weaning approach that promotes babies feeding themselves with appropriate, normal food from four to six months old. If your baby is a good eater, all this will seem easy. If they're fussy, then maybe you'll be making blueberry pancakes for breakfast, lunch and dinner!

Nappies

Oh yes, nappies. Most people use nappies, but they're not essential if you live in a tropical village in the Pacific Islands. Most parents there use the elimination communication method, instead of nappies. This involves no nappies from birth, just timing, signals and baby cues for the caregiver to know when babies have elimination needs. There's a growing movement of parents around the world who use elimination communication. It sounds pretty messy, but those who use it, love it.

If you're using nappies, are they disposable, compostable or cloth nappies for a sustainable approach? Disposable nappies create hundreds of thousands of tons of landfill, taking 200–500 years to break down. The manufacture of nappies typically includes the use of non-renewable plastics and wood pulp. Compostable nappies are great, as long as you find a brand that works for you and a place to safely dispose of them. In contrast, most cloth nappies are made from renewable materials and their biggest environmental impact is likely to be at your home as you wash and dry them. Cloth nappies win. It is more effort for you, but once you've got a system, it isn't too much extra for the return to your bank balance and for the planet. If you're overwhelmed by the volume of cloth nappy washing, consider doing a mix of cloth and disposable nappies.

> ***Living with your baby is an adventure, a journey to someplace new. You choose which fork in the road will take you where you want to go. Leave the rest of the parenting advice behind.***

ECO TOYS

Children love toys! When they're little, they crawl towards them and chew. As they grow up, children begin to build and imagine they're superheroes with the toys and their friends.

Most children have more toys than they can use or store. Their collection is built by well-meaning friends and family over a few years of Christmases, birthday parties and other special occasions. Many toys are plastic, built to last just a few months of play time and destined for a long life in landfill.

It is possible to choose toys that are good for children and don't cost the earth. Consider choosing toys that are:

- **Durable**
  Choose toys that are built to last 20 years, not two months. Consider the workmanship and whether it's up to the rough handling it's likely to get from children.

- **Multi-age**
  Good quality little figures, animals or dinosaurs work for ages one to seven. Duplo and Lego. Yes, they're plastic, but they last forever. Who hasn't seen the vintage sets? Duplo and Lego promote imaginative play as well as fine motor skills in their construction.

- **Multi-use**
  Blocks are brilliant – they're inexpensive and non-branded. Some of the magnetic play sets are also pretty special. In general, the more your child needs to use their mind and body to make a toy work, the more they're learning. So, leave behind those shiny toys in the shopping mall that light up when you press a button.

- **Sustainably manufactured**
  It is possible to buy toys that are certified organic, made from natural, recycled and sustainable materials. You'll find some of these toys in special, local toy stores. There's also a handful of online toy stores that specialise in environmental and ethical toys.

Do you need to own the toys? Many communities have a local toy library. Some are community run, others commercially run and still more operate out of the local book library. Search online or ask your local council to get you in touch with a toy library near you.

WATER WISE HOME

Fresh water is essential for sustaining human life. We drink, cook, wash, garden and play in the fresh water around our home.

Here are seven simply ways to reduce our footprint and be wise with the water that we use.

1. Fix up the leaks

A leaky tap can drip a surprisingly large volume of water, directly into the drain. Fix dripping taps and save up to 60L of water every week. While you're looking at taps, consider making sure your toilet doesn't leak too.

2. Shower with less

Fit your shower with a water-efficient shower head and switch from using 15L/ minute to just 6L/ minute without compromising on quality. While you're in the shower, consider taking shorter showers. Some people like to set a two-minute hourglass timer to make sure they're brief.

3. Better ways to flush the toilet

Old fashioned toilets use 13L per flush. Modern toilets can use as little as 6L per full flush and 4 L per half flush. Not keen to retrofit a new toilet? Simply place a brick (or two) in the top of your toilet cistern to displace the water and 'trick' your old-fashioned toilet into using less water.

If you're super passionate about reducing water use, consider apply the phrase 'if it's yellow, let it mellow' to your toilet habits. In some households, particularly those when tank water is in short supply, every little bit counts and the toilet is not flushed for a wee — the lid is simply closed and flushed when the next household member does a poo.

4. Wash a full load

Save up your washing so you're always washing a full load rather than a few half loads. Washing machines are most efficient when they're full.

5. Water saving appliances

When it's time for a new dishwasher or washing machine, choose your appliance to be water efficient. The difference between a 'regular' dishwasher and a water-efficient dishwasher is about 10 litres per wash. The difference between a 'regular' washing machine and a water-efficient washing machine can be 50 litres per washing load. Choosing water efficient also helps to drive up consumer demand for this feature.

6. Collect rain water for your garden

Install a rain water tank, the largest that will fit. Instead of the rain falling on your roof and going down the drain, you can capture it and use it to water your vegetables or to play in the garden. This could save up to 5,000L in a year.

7. Re-use your grey water

Water from your shower, bathroom sink and laundry is called 'grey' water, because it's only partially contaminated from washing. Most councils have rules about how grey water can be used and it's typically OK to pipe it directly onto a habitat garden or fruit trees. Source a simple grey water system from your local hardware store. If you're keen on a low-tech approach, simply put a bucket in the bottom of your shower and empty it on pot plants. You can also use a bucket to put bath water out on pot plants.

SHARING FOR GOOD

Learning to share is an essential part of childhood. Tears of frustration are cried as children attempt to exert their authority over objects and people. But we all get there… in the end.

Then, we grow up. We get a job and start earning money. If we're lucky… lots of money. That skill of sharing starts to be unlearnt. Almost everything in our consumer society is geared towards individuals. We have our own car, our own mobile phone and our own bread maker in the back of the kitchen cupboard. Every house has a lawnmower, hedge trimmer, leaf blower and whipper snipper.

It's time we went looking for our sharing mojo. I'm sure it's there, just a little buried under the 'convenience' of our modern world. Sharing is a great way to live gently on our planet and build a richer community along the way.

Here are eight ideas to get you sharing for a better world.

1. Join the buy nothing new movement and share your stuff

The world has embraced the *buy nothing new* movement. Since it's humble beginnings in 2013, there are now more than 5 million participants in 44 countries. The *buy nothing new project* supports people to give, share and build community. Have you got some unusual garden tools that you're happy to loan out? Take a photo and let people in your neighbourhood know. Got a glut of garden vegetables? Let people know and they'll collect them from your home.

2. Tools and unusual equipment

Every time you walk into a hardware store, do you walk out with something you didn't intend to buy? Well, hardware stores aren't just about selling stuff for you to own. They're lesser known for hiring out tools and equipment. Next time you've got a building project, do the sums on specialist equipment and sometimes it will be cheaper to hire, rather than buy. You might also like to ask a neighbour or post a 'loan' request on your local buy nothing new group. Lending tools from a neighbour costs less in both time and money.

3. Join your local library for books and toys

Books

There's a wealth of knowledge, adventure and escape between the pages of a book. Local libraries are a fabulous resource for books, magazines and online resources. Many local libraries offer Audiobooks and eBooks to borrow online. The best part about online borrowing is that you don't have to leave your lounge room to borrow, enjoy and return items.

Toys

Are you running out of space to store children's toys? Join your local toy library so that your children can play with different toys each fortnight and you don't have to store them all. Some toy libraries are community-run with members volunteering to facilitate toy borrowing, a great way to connect with a new community.

4. Commercial sharing

Sharing is cool and the world of business has caught onto this one. Uber, Airbnb and Couchsurfing are all great examples of this. There's now a heap of 'ride sharing' platforms available around the world. Can't afford an electric vehicle? Consider using a car sharing scheme.

5. Ideas

Got an idea that you love? Share it! TEDx events are locally run to shine a light on *ideas worth sharing*. There's a diversity, with speakers as diverse as theoretical physics and innovative indigenous business. Ideas don't need to be TEDx worthy to get shared. Join with like-minded people in a book club to share ideas informally or join mind hive, a collaborative, knowledge sharing platform that believes solving the world's problems will come through the sharing of ideas between disciplines.

6. Books and Magazines

Some people have moved into the digital age, buying books and reading them online. Not so for all people. According to the World Economic Forum, more people are buying hard copy books than eBooks. When they're finished, you can pass them onto friends and family to enjoy. This saves space on your book-shelf and you have someone to talk with about the book. Win-win!

7. Community garden

Not enough space for a large vegetable patch? Want to learn about gardening from others? Community gardening is for you. Each family has their own 'plot' in the garden, and everyone contributes to the maintenance of shared spaces. Community gardens are a great way to expand and share your gardening experience.

8. Outdoor adventure toys

The ultimate in adult sharing is with your toys. Do you have a spare mountain bike that you can lend to a friend's family member who's visiting? Are you off to the beach without us? No problem, take our surf boards. Need an extra tent to take an extra person camping? Ask a friend.

Sharing the things we love, to bring joy to others. Now, that's grown up.

NATURAL SKIN CARE

It's cold and windy outside and your skin is starting to look dry and flaky. Eeeek... there's a scratchy, dry skin feeling on elbows and knees. Time to add some moisturiser and give some love — the natural way.

Most commercial moisturisers contain more than 20 ingredients, many with names that are impossible to say out loud unless you have a chemistry degree.

Some of these chemicals are an extract of natural products, like Sodium Cocoyl Glutamate which is derived from coconut oil and fermented sugar. Unfortunately, other chemicals commonly found in moisturiser are harmful to the environment and possibly to you too. Take, for example, parabens, used as a preservative in many moisturisers, known to be an endocrine disrupter.

Commercial moisturisers smell good and feel nice, but if you're natural-beauty-minded you can do better. We don't need a list of 20 ingredients, or large dollops of cream to make our skin soft and smooth. All we need are a few drops of oil.

Cold pressed oil is a natural product, literally squeezed out of its raw ingredient. Think macadamia oil — extracted by pressing macadamia nuts; avocado oil — extracted made by squashing and separating the oil from avocados, and coconut oil — originally made from the 'flesh' inside coconuts. All of these oils are natural beauty products, edible, good for our skin and good for the planet.

Using oil as a moisturiser, rather than a chemist's cocktail, has the advantage of low *beauty miles*. This is because oils are single ingredient products. *Beauty miles*, or the greenhouse footprint of a product, increase as the number of ingredients needed to manufacture the product goes up. *Beauty miles* also increase the further your moisturiser travels.

Choose the oil that's best for you:

All skin types:	macadamia oil, almond oil and cocoa butter.
Young skin:	apricot kernel oil, coconut oil, hemp seed oil, wheat germ oil or olive oil.
Mature skin:	avocado oil or evening primrose oil.

SEVEN SMALL WAYS TO MAKE A DIFFERENCE

Small can be powerful. An ocean is the sum of many billion drops of water. Yet, every drop in the ocean counts… Especially when that drop is part of a ripple of change.

Here are seven small ways to make the switch towards a sustainable future.

1. Switch from pen to pencil

The average pen writes for about 2km, the average pencil for 46km. For every pencil you use you've saved 24 pens! You've also saved on environmental cost of production. Because pens contains ink, plastic and metal their production has a far greater environmental footprint than a pencil. Pencils contain wood and a clay/graphite mixture. If you're super keen on a pen, consider using one that's refillable.

2. Drink loose leaf tea and coffee

Individually wrapped tea bags might last longer in the cupboard, but their wrapping comes at an environmental cost — for both tea and coffee. Many of the 'paper' wrappings are actually lined with plastic, creating waste that isn't recyclable. Some (beautiful) tea bags are made of nylon. Unfortunately, the

nylon and tea mixture goes straight to landfill when you're finished. Choose loose leaf tea. Enjoy it brewed in a pot or with an infuser in a cup.

3. Choose recycled paper

Create demand for recycled product and you're making a contribution to keeping native forests safe from logging. Recycled paper uses up to 50% less energy and 90% less water to produce compared to new paper. Choose recycled paper throughout your home... from paper towels and toilet paper to printer paper.

4. Refill your liquid soap

If liquid soap is your thing then consider refilling the pump packs rather than buying new. The pump pack action is created by a combination of plastic and metal, giving that nice spring. Because of the combination of materials, the top of the pump pack isn't recyclable and goes to landfill. Most pump packs last for years and you can choose refill bottles that are 100% recyclable.

5. Buy local beer and wine

There's been a global resurgence in local craft breweries and distilleries. Be proudly local! Enjoy the taste of your local soil and climate. Support local farmers and entrepreneurs.

6. No junk mail

Say no to printed advertising in your letter box. Do you really enjoy reading it all, or does it go straight to the recycling bin? Make your own note: 'No junk mail please. Community notices accepted.' Think of all those trees that you're saving just by refusing junk mail.

7. Go digital for Movies, Music and TV

An ever-expanding collection of movies, music and TV series creates an ever-expanding pile of stuff for landfill and recycling. There's also that little question of where to store it all... under the bed? Take your old CDs and DVDs to charities for reuse. Scratched CDs and DVDs are, unfortunately, destined for landfill. Their cases can be recycled – just remove the paper labels. Time for a simpler, cleaner digital future.

Choose your drop in the ocean and enjoy being part of the ripple that it creates.

HOW TO MAKE CHANGE THAT LASTS

We all have the power to change the world... our world. The small decisions we make every day can make a big difference. Have you heard of Meat free Mondays? Ever turned your lights off for the annual Earth Hour in March? Have you thought about a buy nothing new challenge? We have the knowledge and technology to live sustainably. The hard part is making a change.

The trick to sustained change is to create a new habit. According to a study by Duke University researchers, more than 40% of actions people perform each day aren't due to active decision making — they're habits. There's plenty of research into changing habits, much of it reporting a set number of days for habits to form, anything from Dr Maltz's classic 21 days to Lally's 66 days. Recent research says there's more to changing a habit than repetition.

So, how can we create a new habit that's good for our lives and good for the planet? Try these three steps for the best chance of success.

Step 1. Set your goal for change

The best goals come from you. They resonate with your heart and your head. Is this goal about impressing someone else, or something you're expected to do? If you own the goal, you're much more likely to follow through and change that habit.

Make your goal realistic in size. Are you decluttering your home? Start with one room each week. Is plastic-free possible or does plastic free for a week suit you better? If you want to grow your own food, start with some pots or one veggie patch — not the full backyard transformation. Want to make your home more sustainable? Take one action each month. Install a water tank in January, replace light bulbs with energy efficient bulbs in February, install water efficient shower heads in March, seal the draughts in April.

Step 2. Trigger your new habit with an existing habit

While your habit is forming, it can be hard to remember to do the new activity. Finding a trigger to help you remember is important, not just a note on the fridge.

If you're getting started growing herbs and vegetables, you'll need to get into the habit of watering (and harvesting) regularly. Maybe you could check your garden each evening, before cooking dinner. Over time, when you start to think about cooking dinner, it will automatically trigger the new habit… a quick tour of the veggie patch. By weaving the new into the old, you're more likely to actually make it through the habit establishment phase.

Perhaps you need to carve out time for sustainable home improvements. Maybe Saturday afternoons, after netball training is your time… every week.

Don't lose hope if you wobble a little bit. It's OK. According to a study published by the *European Journal of Psychology*, missing just one opportunity to repeat the behaviour doesn't matter for habit formation. Start back where you left off and maybe change your trigger or modify your goal.

Step 3. Know what success looks like and track your progress

Halving your household waste is a goal that's pretty easy to measure and track progress. When you take out the rubbish bin each week, you can see what's inside and make notes, if you like.

Tracking improvements in energy efficiency and reduction can be trickier, especially if your power bill comes in quarterly. Get yourself a smart meter or solar panel and battery system that tracks your power usage and power generation, then it is possible to track your power use in real time. Want to know exactly how much electricity it takes to turn on the oven and bake biscuits? It can tell you… exactly.

Staying motivated on your own can be hard work. Friends, colleagues and neighbours can be your best asset here. Track your progress with a buddy to make sure you maintain a good habit. Is there anyone that will match your 'plastic free' challenge?

Changes like *Meat free Monday* and *Earth Hour* start in your home. In every person's home. When everyone changes their habits, we create a new culture and a brighter future for our planet.

HOW TO CHANGE THE WORLD

Greta Thunberg began as an ordinary Swedish student. In August 2018, instead of going to school she began striking outside the Swedish Parliament. Her action was the beginning of a global campaign for stronger action on climate change, involving millions of people. Just one person… changing the future of life on earth.

Any ordinary person can make a difference. Just a small handful of people become internationally influential, like Greta. In a small community, every single person matters simply by expressing their everyday choices through action. Here's how you can make a (bigger) difference and change the world, starting in your community.

1. Start with you

Live your life the way you'd like the world to live. Sound familiar? Mahatma Gandhi is famous for saying, 'be the change that you want to see in the world'. Keen to see farmers being paid a fair price for their produce? Shop at the farmer's market. Wish every person lived carbon positive? Do the paperwork and live carbon positive in your life. People that you know will see what you're doing, even if you don't preach. We're hugely influential on others, simply through our everyday actions and choices.

2. Convert interested people

Ask your family, friends and neighbours if they're interested in knowing more about your passion for sustainable living. If the answer's yes — fantastic, it's time to share your passion. Perhaps it's a tip about soft plastic recycling, 'did you know that soft plastics can be recycled by taking them into supermarkets?' It's a tip that I've shared hundreds of times… which originally came from a friend who knew that I was interested in sustainable living. Enthusiasm is infectious, share your tips with passion and positivity.

The approach that you take to converting interested people is crucial. Don't get trapped by pointing the blame finger or putting them down — this negative approach might work with a few people but it's a total turn-off for almost everyone else.

3. Join with other passionate people

Find a group of people in your local community that share your passion. You'll be able to learn from them, as well as influence action taken by the group.

4. Influence new people

Reach out to people who are new to your ideas. Where can you find them?

- Get social with Facebook, Twitter or Instagram. Tell your online friends what you really think about sustainable living.
- Write a letter to the editor. Newsflash! The letters page in a newspaper is one of the most popular pages. If your letter is published then thousands — yes, thousands — of people will read it, including politicians and policy makers. Not sure where to start with a letter to the editor? Look for practical tips in Communications Guides available on the Australian Climate Council website.
- Start a blog or podcast and build your following. There's a heap of wonderful blogs on sustainability that are written by locals for locals.

5. Influence the government

Write, telephone or meet with your local member of Parliament. It's their job to represent you, so let them know what's important to you! Think local and nationally. You might also want to consider joining a political party. Party members influence the policies that parties develop and take to each election. You can also influence the public service, especially if you keep an eye out for public consultations on sustainable living.

We are all ordinary. It's what we choose to do with our ordinary lives that changes the world and can make an extra-ordinary difference.

GARDENS BIG, SMALL AND RAMBLING

GARDENING BASICS

Veggie patch

As with real estate, think location, location, location. The ideal veggie patch is sunny for most of the day as well as having some shelter from those harsh rays of light on a hot afternoon. North facing sun is best. Shelter from wind is ideal. Create a wind barrier with hedging plants like rosemary, lavender or fijoa trees.

Consider the type of vegetable bed that suits your garden best.

- **Dig directly into the soil. This is the low cost, low fuss option. It's ideal if you already have good quality soil.**
- **Construct a raised bed. The cost of a raised bed is moderate, depending on materials used to create the border and the cost of soil that you buy in. Try to re-purpose materials from around your home for the border and include your own compost in the soil. Raised beds are ideal when your soil quality is poor.**
- **Wicking bed. This is the high cost, high fuss option for construction. However, they're great for lazy gardeners who want lush, juicy vegetables but find it hard to remember to water them. Water in a wicking bed is stored between the stones at the bottom of the bed. It's automatically drawn up into the soil as plants need it.**

Next, make sure you've got good quality soil. Vegetables like their soil nutrient rich and many will grow out of a compost heap. Every season add a combination of animal manure, compost and worm casings to your vegetable beds. Consider enriching a 2m x 1.5m patch with a full 25 litre bag of cow or sheep manure and four buckets of compost. Maintain your soil by using plenty of mulch and a little seaweed liquid fertiliser each month.

The final foundation is water. Before you plant, know how you're going to get water to your vegetables. If you're time poor, try setting up a wicking bed or a drip irrigation system with a timer. Vegetables like the soil to be moist. Generally, they won't grow as well when the soil is too dry or waterlogged.

If you're planting seeds, prevent slugs and snails from flattening your fledgling crop overnight with broken-up egg shells, sawdust or pet-friendly anti-slug pellets.

Productive pots

If there's light, soil and water you can grow your own food. Pots on a balcony, windowsill or on a vertical wall are perfect for growing food in a convenient location. There's almost no limit to the vegetables and dwarf fruit trees that you can grow in pots.

Finding the right place for your pots will depend on the herbs, vegetables or fruit trees that you wish to grow. North facing, with plenty of light, is generally best. West facing areas tend to overheat in summer and might need shading from the harsh afternoon sun.

The best thing about growing food in pots is that you can move them around. In the summertime, move the pots into shade. In the wintertime, protect them from frost.

Fill your pots with soil that is rich and sandy, providing your plants with plenty of nutrients and without waterlogging. You can make your own 'potting mix' by mixing compost, worm casting, river sand and a little garden soil.

Remembering to water is crucial for plants in pots. Keeping pots near to the kitchen or beside a regular pathway is perfect for first time gardeners. Water seedlings every day and then every second day after plants are established.

Fruit trees and more

With space for a larger garden, use the principles of permaculture to guide your set-up. Grow herbs and vegetables that you use regularly in cooking close to your home. Have an orchard beyond the veggie patch, along with a chicken run and a beehive. Your compost and worm farm can go somewhere between the vegetables and the orchard. Beyond the orchard consider planting native plants as habitat and food for native birds.

MICROGREENS

Microgreens are essential to creating the beauty that we love in restaurant food. They're tiny shoots from baby vegetables, harvested a few weeks after they emerge from the soil and sprinkled on as a gorgeous garnish.

As well as looking good, research has shown that they're also packed with nutrients. The United States Department of Nutrition and Food Science have published studies on microgreen nutrients. All of these studies have found that microgreens contain nutrient densities that are between four and forty times higher than mature leaves.

Here's how you can get the beauty of a restaurant meal, with a burst of nutrition, by growing your own microgreens at home.

Step 1. Choose a sunny spot

Preferably near the kitchen window. You'll want to enjoy these mini plants as they sprout out of the soil and grow. If they're near the kitchen, this will help you remember to use and water them.

Step 2. Find a pot

The pot doesn't need to be deep, as you'll be harvesting the microgreens before roots need a lot of space. You can make a pot at home by re-using large takeaway containers. Simply pierce a few holes in the bottom so the seedlings don't get waterlogged and place them on top of an old plate.

Step 3. Fill up your pot with rich soil or potting mix

Steer clear of fresh manure so that you don't have to worry about it contaminating your kitchen.

Step 4. Choose your microgreen seeds

You can grow almost any type of vegetable seedling, depending on your taste and desire for colour. If you have seeds that need using in the back of your gardening cupboard, start with them. Here are some popular microgreens that you might like to try:

- Red cabbage and beetroot — lovely rich colour
- Coriander and basil — aromatic flavour
- Mustard, cress and rocket (arugula) — peppery flavour
- Broccoli, cabbage, kale, cauliflower — healthy goodness
- Sunflower and chia — nutty flavour

Step 5. Plant your seeds

Keep the soil in your pot moist by watering every day or two. Because the seedlings are delicate when they first emerge, you might find a spray bottle is a good way to keep emerging seedlings moist.

Step 6. Harvest!

When the seedlings are about two weeks old and pushing up their 'second leaf' it's time to harvest. Not sure if you've got the harvest time right? 'Second leaf' refers to the second type of leaf that emerges from your seedling. It's got true characteristics of the mature plant, so the second leaf for red cabbage is actually red and not green like the first leaf.

Harvest with scissors by gently snipping off the stem above the soil. Generally, you'll get one harvest for each seedling. After you've harvested the whole crop, it's time to refill the pot and start again!

HERBS

Fresh herbs turn an ordinary dish into a flavour sensation. In the oven, roast rosemary with potatoes, crispy sage with pumpkin and parsley on carrots. In fresh salads, basil pairs perfectly with tomato and mozzarella cheese and Asian basil, mint and coriander complete the Vietnamese rice paper roll flavour. Yum.

The best herbs to grow are the ones that you like to eat. The more you harvest, the more new leaves will sprout. Most herbs grow well in pots, vertical gardens and inside your veggie bed.

Annual herbs

Herbs that grow in the spring, summer and autumn, then need to be grown again from seed the following year.

Basil

Easy to grow from seed or from small plants, basil is a summertime annual. The best time for planting is spring. Plants are ready to harvest when the stalks are at least 15cm tall. Continue harvesting until your plant has flowered. Basil dies in cold weather and is wilted by frost. Basil needs full sun, rich soil and regular watering.

There are three main varieties of basil. Sweet basil is used in Italian cooking — think fresh tomato, basil and mozzarella salad. Thai basil is perfect for Asian stir-fries and rice paper rolls. Holy basil (Tulsi) is used in Indian cooking and as a popular herbal tea.

Coriander

Easy to grow from seed in the autumn, coriander is a cooler weather annual. The best time for planting is autumn. However, it can be planted throughout spring and summer into a shaded garden. Coriander is fickle and responds to heat stress by producing flowers, not the leaves that we like to eat. Common causes of stress to coriander in the home garden are a lack of water or transplanting of seedlings. Coriander prefers full sun or part-shade, rich soil and regular watering. Because it crops best throughout the wintertime, consider planting it in the same veggie bed as garlic.

Coriander is an essential Asian and Indian flavour — essential in a Thai stir fry and a perfect garnish for lentil dahl.

Dill

Easy to grow from seed, dill loves warm (but not hot) weather, a frost-free location and rich, well-drained soil. The best time for planting is spring. Plants are ready to harvest as soon as the fronds are 5cm long. Continue harvesting until your plant has flowered. Keep the flowers to attract aphids and other helpful insects to your garden.

Dill creates a subtle yet intense flavour. Stir it through sauteed leafy greens or top your yoghurt and beetroot salad with finely chopped dill leaves. Small quantities can make a big impression.

Parsley

Parsley grows well with almost any amount of neglect, sunshine, shade, drought or flood. Parsley grows easily from seed or from transplanted small plants. Parsley continues to grow year-round. It self-seeds, so if you leave the flowers, you'll have parsley popping up all over your garden. Parsley is biannual. In the right location it will grow and produce parsley for two years, instead of the regular single year of the annuals. There are two main types of parsley: flat leafed (Italian) parsley and curly leafed parsley.

Use parsley leaves fresh on top of salads, tossed through boiled potatoes or in a green chilli salsa.

Rocket (Aragula)

Rocket will grow (almost) anywhere. While it prefers full sunshine, water and rich soil, it will tolerate almost any conditions. Grow your rocket from seed, its small seeds are one of the fastest to germinate. Rocket continues to grow year-round but prefers the cooler months as it goes to seed in the summer.
It also self-seeds, so if you leave the flowers, you'll attract bees to your garden and will have rocket popping up everywhere next springtime.

Fresh, peppery and edgy in any salad. Rocket is also great for making pesto, just substitute it for basil.

Perennial herbs

Herbs that grow from cuttings or runners which die back in the wintertime and sprout again from their roots in the springtime. Some of these herbs are deep rooted, like comfrey and rhubarb.

Comfrey

Grow your comfrey from seed, from a small plant in a pot or ask a friend with a large plant to dig up a clump. Comfrey grows well in full sun or part shade, with rich soil and plenty of water. It dies back in cold weather. If it likes the conditions, comfrey will take over.

Comfrey isn't grown for eating — the leaves are poisonous. It's grown as a garden fertiliser and compost 'accelerator'. Because of its deep roots, comfrey draws nutrients into its leaves. Harvest the leaves and put them into your compost or around the base of fruit trees as a natural fertiliser.

Mint

Get your mint in a small pot or ask a friend with a large mint plant to dig up a runner. Mint grows well in full sun or part shade, with rich soil and plenty of water. It dies back in cold weather and can be fussy. However, if it likes the conditions, mint will take over. Most people prefer to leave mint out of their veggie patch, or create an entire patch filled with mint.

Mint is an essential savoury Moroccan flavour; it gives brilliant zing scattered on top of salads and is the perfect garnish on a watermelon salad.

Oregano

Oregano grows well from seed or from runners. Ask a friend with oregano in their garden to dig up a runner for you. It prefers full sun, rich soil and frequent watering, but will grow with almost any amount of neglect. Oregano is best in the spring, summer and autumn times. It dies back if the winter is cold.

Use oregano in Italian and Greek cooking. It's essential to creating an authentic Italian tomato sauce for pizza or pasta.

Rhubarb

Once you've got rhubarb established in your garden, you'll have it forever. It's a herb that grows long red or green stems and large, dark green leaves from crowns in the soil. Looking for a new rhubarb plant? Ask a friend with a healthy plant to split their crown in late winter and pop your crown into rich, moist soil. Voilà — rhubarb and deliciousness for life.

In cold climates, rhubarb dies back in winter. So, rather than waste the stems, harvest them each autumn.

Rosemary

Originating on the rocky hillsides of the Mediterranean, rosemary is a very hardy plant. It prefers full sun, rich soil and frequent watering, but will tolerate almost anything except a lack of sunlight. Grow rosemary from a small plant in a pot or ask a friend for a few cuttings and see if these can take root by leaving them in water on a windowsill for a few weeks. Rosemary grows year-round.

Use rosemary with lamb and potatoes in the winter and in a green salsa in the summer.

Sage

Typically grown from small plants, sage can also be grown from seed. It prefers full sun, rich soil and regular watering and will tolerate almost any amount of neglect if there's sunshine and the roots aren't waterlogged. Sage is best in the spring, summer and autumn as it loses some leaves when there's a frost. If you have a large enough plant, you can have sage year-round in all climates.

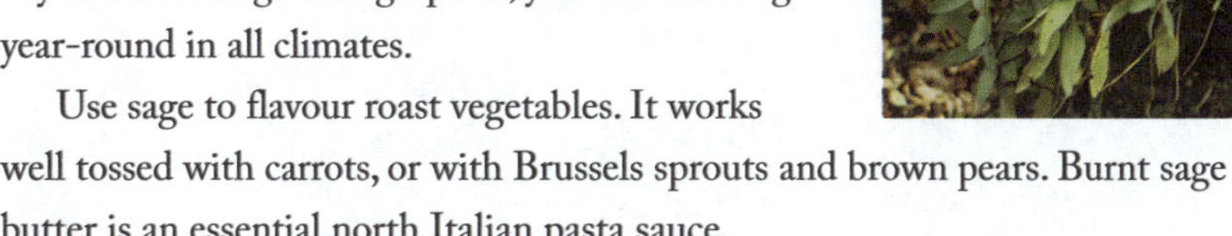

Use sage to flavour roast vegetables. It works well tossed with carrots, or with Brussels sprouts and brown pears. Burnt sage butter is an essential north Italian pasta sauce.

Tarragon

Tarragon grows well from seed or from runners. It prefers full sun, rich soil and frequent watering, but will grow with almost any amount of neglect. Tarragon is best in the spring, summer and autumn times. It dies back if the winter is cold.

Use tarragon in Middle Eastern cooking. When fresh, it has an aniseed-like zing and is gorgeous in a light, green salad. Try tossing carrots in fresh tarragon and roasting them or using tarragon to flavour chicken or fish.

SPRING VEGGIE PATCH

Green shoots transform the garden in spring. Asparagus spears push up through the soil, daring us to snap off the sweet and tender shoots. White apple blossom flies into the air with each gust of wind, creating a majestic carpet underneath the trees. After a slow and cold winter, everything is growing now... and fast.

Artichoke (Globe): Easy ★☆☆

Globe artichokes are related to thistles. Harvest your artichoke flower in the springtime when it looks like a large, green bud. If you leave the artichoke to flower, it sprouts majestic purple fronds.

Artichokes are most easily grown from small plants or suckers that have been separated from the 'mother' plant in early spring. They're incredibly hardy plants and reasonably drought tolerant. While they prefer full sun and rich soil, they'll grow in almost any conditions. Each plant produces 2–3 chokes (flowers) per year.

Artichokes are perennials, so they need a permanent location in your veggie patch. They sprout in spring, grow about 1.5m wide and 2m tall, then die back in the late summer after flowering.

Eat young artichoke flowers whole. They're gorgeous lightly sautéed with olive oil and herbs with a generous squeeze of lemon juice. More mature flowers can be steamed and drizzled with olive oil and lemon juice, then each flower 'petal' can be individually eaten.

Artichoke (Jerusalem): Easy ★☆☆

Jerusalem artichokes are grown for their tubas, like potatoes or parsnips. They're incredibly hardy and will grow in almost any soil type with full sun or part shade. Set aside a permanent space for artichokes as they'll continue to pop up unless all of the tubas are harvested. They're actually a type of sunflower and a patch of Jerusalem artichokes will produce both lovely, tall sun flowers and edible tubas.

Plant Jerusalem artichoke tubas in early spring. You can source tubas from both garden stores and from the supermarket shelf (vegetable section). They'll sprout and become tall sunflowers by mid-summer. In autumn, the sunflowers die back and it's time to harvest the tubas by 'bandicooting'— taking the tubas that you need for each meal at a time.

Jerusalem artichokes have a delicious nutty flavour and are lovely oven-roasted or in soups. Use them in the same way that you'd cook and prepare a potato. While they're delicious to eat, most people produce distinctively smelly farts after eating Jerusalem artichokes.

Asparagus: Hard ★★★

Asparagus needs a permanent position in the garden with full sun and an annual dose of manure. The asparagus spears that we love to eat emerge from 'crowns' in the ground each spring.

Asparagus is typically grown from *crowns* or roots that are planted 15–20cm deep into a nutrient-rich veggie bed. Plant additional crowns in a 30–50cm spacing along a row. After planting it's time to play the long game. In their first two years, asparagus shoots (or spears) are too thin to be harvested. Let these spears grow into ferns, to strengthen the underground crowns. After waiting two years, you'll be rewarded with lovely, thick spears that emerge from your soil each spring for 30–35 years. Most crowns will produce one or two spears per week in the springtime. To have enough spears for a fresh family meal each week, plant at least 10 asparagus crowns. Be vigilant, spears grow at least 2cm each day and if you're not careful, the spear gets too big and becomes tough and woody. After you've grown tired of eating asparagus, let the spears develop into ferns in the late summer. The ferns will die back over winter, so the spears sprout again from bare earth in early spring.

Harvest asparagus by cutting off the spears a few cm from the ground, when they're at least as thick as a pencil and 15–20cm tall. Lightly sauté in olive oil then toss through a squeeze of lemon juice and pinch of salt.

Beans (Climbing): Easy ★☆☆

Stringless climbing beans are easy to grow and delicious to eat. Choose from the classic green beans, like blue lake climbing beans, or heirloom varieties of purple or yellow stringless climbing beans.

Beans love the sun and rich soil, but they'll grow almost anywhere. Plant seeds directly into the soil, every 15–20cm, below your 2m high climbing frame. Beans are frost sensitive — sow seeds after the risk of frosts have passed and when the soil is warm. Pick beans regularly to stimulate more flowers and beans.

Harvest beans when they're about the length of your hand and eat them directly off the vine or steamed with a dollop of butter and mustard.

Children's tepee made of beans

A great joint project for kids and parents, it's fun and nutritious at the same time. Find a part of your veggie patch that's got plenty of sun, rich soil and where you're happy for the kids to frolic. Mark out enough space on the ground in a circle, for the bottom of the tepee. The space needs to be large enough for two small children to sit down.

Now, it's time for the tepee frame. Source at least 6 long, straight sticks – to create the tepee frame. Use sticks that are about 1.5m long, left over from the winter fruit tree pruning. You can also buy purpose-created sticks from garden stores, made from bamboo or tea trees. Secure the tepee frame by pushing the sticks gently into the garden bed and securely tying the top of the sticks to each other with some string. Secure the sticks in a tepee shape by looping string around each stick and connecting them too each other at about 30cm intervals along each stick. The largest looped string will be at the bottom, giving the beans extra scaffolding to grow upon.

Finally, planting time! Near the base of each stick, plant two bean seeds — one on either side of the stick. Our favourite variety is blue lake climbing beans as they're plentiful and stringless. The tepee works equally well with any variety of climbing bean or even with snow peas. Just make sure that you haven't selected a dwarf variety as these may not climb high enough to cover your tepee.

Now you can watch the bean seedlings grow and create a gorgeous green covering around the tepee frame. About nine weeks after the original planting, the tepee will start to flower and 10–12 weeks later there'll be beans to eat. A tepee made of beans is a gorgeous garden feature; it's a place for children to play and a source of delicious, fresh beans to eat.

Beetroots: Medium ★★☆

Beetroots come with incredible colour. There's the classic purple beetroot available in most grocery stores as well as delicious yellow heirloom varieties. When to plant depends on your climate. In cooler climates plant beetroot seeds or seedlings from spring through to early autumn, in warm climate beetroot seeds and seedlings can be planted almost year-around except in the heat of summer.

Plant seeds directly into rich, well-drained soil, in rows and with full sun. When the seeds are 3–5cm tall, thin the seedlings out and use the small beetroot leaves as a microgreen garnish. Alternatively, plant seedlings into your veggie bed with about 5cm between each plant. Beetroot take at least 8 weeks to grow the delicious beet. Keep plants well hydrated and you'll have sweeter beets.

Eat beetroots both fresh — delicious as part of a grated 'rainbow' salad — or oven-roasted.

Harvesting beetroot leaves is the perfect way to get double value from this delicious root crop. If you're growing beetroot for leaves, enhance a regular vegetable soil with just a little extra manure.

Carrots: Medium ★★☆

Full sun and well-drained, loose soil create the best conditions for carrots. Couple this with regular watering and you'll have sweet, crunchy carrots direct from the garden. Warm climates can plant carrots year-round and cooler climates plant just in the spring and summer time. Plant loose seeds or seeds set in tape in rows. Thin carrot seedlings to be 3–5cm apart. You can eat the sweet and crunchy mini carrots as you thin them out.

Carrots like soil that's easy for their roots to push through. If you have clay soil, loosen it up with a lighter, imported soil or sand. Too much fertiliser can split the carrot roots, just home compost and worm wee is perfect for carrots. Keep your carrots in the ground and harvest them as you're ready to eat. Legend has it that carrots get sweeter if they're in the soil after a frost.

Carrots are delicious fresh in salads or roasted in the oven. Carrot tops are also edible and can be chopped into sautéed greens or pesto — the bonus of growing carrots at home.

Celery: Hard ★★★

The key to producing celery with that delicious juicy crunch is regular watering and keeping the slugs and weavels away. That's in addition to full sun and rich soil. Successful celery requires a whole lot of love.

Plant celery seeds into trays, then transplant these into your garden when the seedlings are about 5cm tall. Give seedlings a 20cm space between plants. Monitor regularly to make sure the ground doesn't dry out and the slugs haven't set up camp.

Harvest home celery stalk by stalk — just take what you need for each meal. That way, you'll have plenty of celery with just a few plants.

Celery is delicious fresh, especially as a healthy snack with peanut butter. Celery tops are also edible and create beautiful flavour as part of a vegetable soup's sofrito.

Cucumber: Medium ★★☆

Cucumber vines ramble towards the sun and they're best grown on a 1.5m high trellis. Grow the type of cucumber that you most prefer to eat: Lebanese, apple, burpless or continental. If you grow more than you can eat, make cucumber pickle.

Plant seedlings or seeds directly into your veggie bed. Rich soil and full sun create the conditions for cucumbers to thrive. However, they'll survive and produce tasty cucumbers in part-sun.

Lettuce: Medium ★★☆

Freckles, Australian yellow leaf, cos, little gems… all types of lettuce. The most forgiving for home gardening would have to be Australian yellow leaf. Others can go bitter at just the slightest hint of dry soil. Lettuce likes rich soil, plenty of water and full sun. Most varieties will tolerate part shade.

Scatter a whole row of lettuce seeds, so that you can harvest leaves when they're young and tender. Most lettuce varieties grow just fine when they're close together. If things get too tight, thin out your row. Lettuce is an annual, producing flowers (instead of leaves) in late spring.

Snails, slugs and lettuce have a special relationship. There are organic options for snail and slug control, and crushed eggshells are a home-grown option. However, the most effective pest control for lettuce is to pick your own, regularly. That way you can pick off slugs individually by looking between leaves… and feed them to your chickens.

Lettuce can be grown year-round if your garden doesn't have any frost.

Potatoes: Hard ★★★

Home grown potatoes have a unique crunch and flavour that makes store-purchased potatoes seem like an entirely different vegetable. Experiment with the enormous variety of potatoes that can be grown at home, from the better-known varieties of Pontiac, Kipfer and Desiree to the lesser-known King Edward and Royal Blues. Potatoes are suited best to cool climates. Warmer climates are better suited for sweet potato.

Potatoes need a special space to grow, as you'll need to heap them up to create maximum yield. Plant seed potatoes into loose, well-drained soil in a position that has full sun. Plant seed potatoes in early spring, after the risk of frost has passed. As the potato plant pokes its leaves up, cover them with a combination of compost and lawn clippings or soil — leave just a small part of the plant poking out of the top. Repeat as often as the potato grows up through its covering. Each time you add this layer of compost and soil, more potato tubers will grow for you to harvest. Some people create a more structured mound, by placing wire with holes large enough to fit your hands through for harvesting along the sides of each row.

You'll know the potatoes are ready to be harvested when the plant has flowered and begins to die back in late summer. Usually, potatoes are best left in the ground and harvested when you need them by 'bandicooting'. If you've got potatoes in the ground, watch to make sure the soil is well-drained. If the potatoes get waterlogged, they will rot.

Home grown potatoes are delicious scrubbed with their skin on, boiled in a pot and served with a little salt and herbed butter.

Snow peas: Easy ★☆☆

Prepare a 2m high climbing frame for your snow peas. Place it in a vegetable bed with rich soil and plenty of sun, then plant seeds along the row with your climbing frame. Water regularly. Because they've got nitrogen-fixing nodules on their roots, snow peas will improve the soil for you.

Fresh snow peas have a beautiful crunch and snow pea shoots add 'wow' factor to any spring salad. Snow peas planted in spring will be ready to harvest in early summer, and some plants continue producing until autumn.

Sugar snap peas: Easy ★☆☆

Prepare a 1.5m high climbing frame for your sugar peas. Place it in a vegetable bed with rich soil and plenty of sun, then plant seeds along the row with your climbing frame. Water regularly. Because they've got nitrogen-fixing nodules on their roots, sugar snap peas will improve the soil for you.

Sugar snap peas look and grow like regular peas. That's where the similarity ends. To eat, they're sweet and the shells are tender. Eat the whole sugar snap pea in salads, lightly steam as a side vegetable or lightly cooked in a stir fry. Sugar snap peas planted in spring will be ready to harvest in early summer, and some plants will continue producing until autumn.

Sweet potatoes: Medium ★★☆

Hello colour, hello flavour... sweet potato has arrived. While sweet potatoes can be grown in any climate, they prefer warm weather and vines are frost sensitive.

Plant tubers or cuttings in the springtime, after the risk of frost has passed. Sweet potatoes like full sun but will tolerate part-shade and they like rich, well-drained soil. The vines grow along the ground and will climb up a frame, but the real action is underground. In autumn, when the leaves die back, it's time to harvest. Because it can be difficult to dig every sweet potato out of the ground, they're best grown in a dedicated garden patch.

Radishes (globe): Easy ★☆☆

Crunchy, peppery and juicy. Radishes add something special to salads. Growing your own will mean that you don't have to throw out unused radishes from a large, supermarket bunch. Just pick what you need — fresh every time!

Globe radishes are very forgiving plants. They prefer a rich soil, full sun and regular watering. They also seem to grow just fine, with poorer soil, dappled shade and neglectful watering. However, for a crop of radishes that are sweet and peppery, rather than just peppery, regular water is essential.

Herbs in the veggie patch

To supply an active home kitchen with plenty of herbs, grow your annual herbs along a row inside your veggie patch. Herbs like the rich soil, full sunlight and regular water that you'll be supplying vegetables. Putting herbs in your veggie patch gives you space that will enable you to generously use them, meal after meal. Sprinkle half a packet of seed, along a 1–1.5m long row and thin out the seedlings. The best herbs to grow inside your veggie patch are parsley, rocket, basil, dill and coriander.

EDIBLE FLOWERS IN THE VEGGIE PATCH

Spring blooms delight our senses with many colours, shapes and smells. This marks the beginning of a productive summer veggie patch. Flowers are a sumptuous display of plants desperately seeking their true lover. It's no coincidence that spring is the season for new lovers.

Make the most of spring's celebration of love by planting edible flowers in your veggie patch. Here are some of the best:

1. Pansies

These gorgeous little flowers are typically grown in long, cold winters. The right variety will continue growing (and flowering) throughout spring and summer as well. Many pansies have an exquisite aroma. They add flair to almost any kitchen dish. Pansies are a traditional English decoration on top of cakes. Just dip them in egg white and then in caster sugar. They're also a very pretty addition to crepes. Place them on top of your crepe while the first side is cooking. Let them cook into the crepes while the second side cooks.

2. Giant sunflowers

These are a true summer crop and a spectacular addition to any veggie patch. Giant sunflowers add wow factor, as they grow to the height of a human adult. After the bees have enjoyed your sunflower, birds will enjoy the sunflower seeds… unless you pick them first. You can eat the seeds and the flower petals. Plant the seeds of your giant sunflower in the springtime.

3. Marigolds

These brightly coloured blooms are all-round brilliant. Not only are they edible and known in some parts of the world as poor-man's saffron, they're also great for repelling garden pests. They're best planted around the border of any veggie patch. Marigolds don't tolerate frosts, so they need to be planted after the frost (or with frost protection). In the autumn time, if you let marigolds go to seed, you'll have a mass of seeds that you can store over the winter then plant again in the springtime. You can use both marigold flower petals and the spicier leaves in a summer potato salad. Try the age-old tradition and use marigold leaves in a stew, to give it a lovely orange colour.

4. Nasturtiums

Delicate, bright and sweet. Nasturtium flowers have a long, thin tube that protrudes below the flower petals and contains nectar. Snapping off this thin tube and sucking out the nectar is a favourite activity for children. Nasturtium flowers add colour to salads and can be frozen in the middle of fruit juice ice blocks — transforming a humble, homemade ice block into something truly special. Plant nasturtium seeds in the springtime and protect them from any late frosts.

5. Let your winter crops bloom

Most gardeners try to stop their winter crops from blooming, so the plants focus their energy on their prized leaves. Instead, let some of your winter crops bloom. Kale and broccoli will produce a mass of gorgeous yellow flowers. Rocket and coriander both have delicate white flowers. Bees just love these little flowers. They're also a fantastic addition to a leafy green salad. Just toss an equal amount of baby spinach and blooms from your winter crops together, top with toasted seeds and dress with extra virgin olive oil and vinegar.

SUMMER VEGGIE PATCH

Long days of sunshine. Green shoots. Earthworms dancing in the soil. Share that summertime buzz by getting your vegetable garden growing with these sun loving vegetables.

These veggies can be planted as seeds in spring, but they live for the warmth of summer and should be transplanted into the veggie patch in late spring or early summer. They thrive on hot summer days in the veggie patch.

Capsicum (Bell peppers): Medium ★★☆

Warmth and plenty of sunshine are essential to developing sweetness in capsicums. Originally from South America, capsicums are better known for their role in Mediterranean dishes like Spanish *paella.* While large, red bell capsicums are most common, other varieties of capsicum are just as easy to grow, including 'chocolate' capsicum that's a deep purple colour and yellow capsicums. Green capsicums are simply unripe capsicums of any variety.

Full sun, regular watering and rich, well-drained soil are important for capsicum plants. Plant seeds into pots and transplant into the veggie patch when the seedlings are about 5cm tall. Plants grow into small, self-supporting shrubs. Capsicums can be grown year-around in warm climates. In cool climates, grow your capsicums in spring and summer — avoiding cooler weather and frosts.

Harvest when your capsicums are the size and colour that you like to eat. Raw capsicum is delicious in salads. Char grill whole or halved capsicums, let them cool and remove the skin to create delicious roast capsicum strips for an antipasto.

Eggplant: Medium ★★☆

Eggplants, also known as aubergines, have a delicious, rich flavour that can be very filling. Most people prepare eggplant by slicing and salting the flesh, then washing off the salt before sautéing or roasting. However, eggplants can be roasted or chargrilled whole. Eggplant varieties include the common, large sized 'black night' with black skin. There are also long, thin eggplants with light purple skin, including 'little fingers' or 'ping tung long' that are used more commonly in Asian cooking.

Full sun, regular watering and rich, well-drained soil are important for eggplants. Plant seeds into pots and transplant into the veggie patch when the seedlings are about 5cm tall. Plants grow into small, self-supporting shrubs. Eggplants can be grown year-round in warm climates. In cool climates, grow your eggplants in spring and summer — avoiding cooler weather and frosts.

Harvest when your eggplants are ripe and the fruit is firm but has a little 'give' when you lightly squeeze.

Leeks: Easy ★☆☆

A row of fresh leeks in the garden creates a fabulous versatility in the kitchen. Leeks can be harvested when they're mature and used in soups or sautés. Baby leeks can be harvested to thin out a row of leeks and they're delicious roasted or grilled with a little butter.

Plant leek seeds along a row with full sun and rich soil. As the leeks grow, thin them out so that each plant has enough space to grow and to make sure future harvesting doesn't disturb the roots too much.

Melons: Hard ★★★

Watermelon, Rock Melon and Honey Dew Melon all grow from vines that thrive in hot weather.

Plant seeds directly into rows and thin out smaller seedlings. Alternatively, plant seeds into seedling trays and transplant into your veggie bed when plants are about 5cm tall. Melons like rich soil, full sun and regular watering — not water logging.

In warm climates, melons can be grown year-round. In cool climates, seedlings should be planted out in spring as soon as the risk of frost has passed. Melons need to be planted as early as possible because it takes a few months for the vine to mature and for melons to ripen — this is part of what makes melons tricky in cool climates. You'll know that your melon is ready when it sounds hollow if you tap it. You'll also notice that the vine directly connecting to the melon will start to shrivel. As the melons grow, lift the fruit off the soil and place it on a bed of straw to reduce the risk of the fruit spoiling.

Pumpkins: Easy ★☆☆

Pumpkin vines send their tentacles towards sunshine, including climbing up trees. Unchecked pumpkins create a lush, rambling garden vibe. They can also take over your summer crops, so pay careful attention to pumpkin vines. Choose your pumpkin variety based on the type you like to eat. Butternuts are smaller and mature faster, but the hard skin of Queensland Blue pumpkins makes them a hardy garden favourite.

Plant seeds directly into rows and thin out smaller seedlings. Alternatively, plant seeds into seedling trays and transplant into your veggie bed when plants are about 5cm tall. Pumpkins like rich soil, full sun and regular watering. However, they'll survive with almost total neglect. Many people find that productive pumpkin vines will grow out of the compost heap.

Pumpkins are ready for harvest when the stem changes colour and goes hard. Pull it off the vine and pop it upside down in the sun for a day or two to harden the skin for storage.

In warm climates, pumpkins can be grown year-round. In cool climates, seedlings should be planted out in spring as soon as the risk of frost has passed.

Sweet Corn: Medium ★★☆

Sweet corn that's home grown has a unique sweetness and richness that's best experienced when a sweet corn ear has been picked, husked and eaten fresh in the veggie patch. For the uninitiated, it's an out of world experience.

Sweet corn takes quite a bit of space as you'll need to have at least 20 plants, preferably across a few rows, to enable effective pollination. Corn needs rich, well-drained soil and full sun. Plant seeds directly into the soil, with two seeds every 20cm along a row (or in a square pattern). When seedlings are about 15cm tall, thin out the smaller plants.

Corn is ready to be harvested when the tassels at the end of each corn ear turns from yellow to light brown. When you remove the husk, each of the fleshy corn seeds should be well formed along the cob. Harvest corn as close as possible to the time when you wish to eat it because it will lose flavour quickly.

Prepare your corn cob for eating by steaming then tossing the corn in salt and herb butter.

Tomatoes: Easy - Medium ★☆☆

So many varieties, pick your favourites to grow at home. Experiment with lesser-known varieties for unique flavour and their *wow* factor, like black Russians or yellow honeybees. Easiest to grow, by far, are the smaller varieties like Tiny Toms, Yellow Grape or Cherry tomatoes. The larger varieties need their limbs tied to stakes, so that the huge tomatoes can be supported as they ripen.

Tomatoes like rich soil and lots of sun. They're easy going when it comes to watering, preferring moist soil and coping alright if you forget to water for up to a week. They prefer to be underwatered when fruiting. Heavy rain will split tomato fruit.

Plant tomato seeds into small pots and transfer them into the garden or a large pot. In warm climates, tomatoes can be grown year-round. In cool climates, seedlings should be planted out in spring, as soon as the risk of frost has passed. Tomatoes can become diseased if you grow them more than two years in a row in the same pot or vegetable bed so move them around your garden.

Zucchini: Easy ★☆☆

Grown at home you can pick zucchinis at any size. Small zucchinis are sweet and tender, perfect to be eaten raw in salads or spiralised as a pasta alternative. Large zucchinis are great for cooking. Think zucchini fritters, chutney or cake. If you're a gourmet cook, fresh zucchini flowers are a big reward for growing your own plants at home. You can also grow different zucchini varieties, including yellow or light green coloured zucchinis.

Zucchinis like rich soil and full sun. Regular watering is preferable, but they'll tolerate quite a bit of neglect. Zucchinis can be grown in large pots, but they grow best with the space that comes with a vegetable bed. Plant seeds into small pots and transfer them into the garden or a large pot. If you plant seeds directly into your veggie bed, thin your seedlings to prevent problems with mildew.

AUTUMN IN THE VEGGIE PATCH

It's time for a change of pace, from the heat and excess of summer to a gentle and steady pace. The secret to a productive winter garden in cool climates is to plant veggies early in autumn, when the days are long and the soil is warm. This often means taking out productive plants to make way for the seedlings of autumn and winter veggies. If you wait until late autumn, when the first frost destroys your tomato leaves and pumpkin vines, then you will have missed the best growing days.

Brassicas: Hard ★★★

Kale, bok choy, wombok, broccoli and rocket are all part of the same brassica family. These beautiful plants all like soil that has been enriched with well-rotted manure or compost. The same amount of manure to garden space applies for brassicas as it applies to spinach. Brassicas do best with plenty of sunshine and regular water.

Autumn is the right time to transplant brassica seedlings. You'll want seedlings for most brassicas to be about 30cm apart, as each plant needs space to prevent aphid attack. Expect your first harvest about 10 weeks after you transplant the seedlings.

Most people who grow brassicas are familiar with small, white moths that like to share in the produce. White moths, and holes in your brassica leaves mean that it's time to act – you've got an infestation of cabbage moth caterpillars. There's an organic, biological control for these caterpillars called Dipel.

Young brassicas are also affected by snails, which can eat several plants in one night – don't be fooled into thinking you have a possum problem if you haven't ruled out snails. Older brassicas can be affected by scale, especially when they're tightly planted. Try to isolate the scale by removing affected leaves, and use an organic, white oil spray.

Broccoli

Broccoli cut freshly from your garden tastes sweeter than broccoli that's been picked, cold stored and transported to a supermarket shelf. If you haven't already, try it and you'll be hooked forever on the good home-grown stuff.

Try sprouting broccoli for a long, steady harvest period. You can get purple as well as plain green. While spectacular in the garden, purple broccoli they look the same as green broccoli when they're cooked. Sprouting broccoli doesn't form one large head; it produces many heads that will give you a longer harvest.

English spinach: Medium ★★☆

Baby spinach is so tasty, so easily ruined in the back of the fridge… and so easy to grow! As a bonus, you can harvest leaves both early and late to have both baby spinach and regular spinach available throughout autumn and winter. These gorgeous little plants like soil that has been enriched with well-rotted manure or compost. Spinach does best with plenty of sunshine and regular water.

You can plant seeds or transplant seedlings, depending on how keen you are to make that harvest. Get seeds into the soil as soon as possible, to give them a head start before winter establishes. They're annuals, so spinach will produce leaves over the winter and then bolt and flower as soon as the weather gets warm in late spring.

Young plants are a delicacy for snails. Consider an organic remedy from the garden store or crush up eggshells and put them around your plants. The crushed eggshells are difficult for soft snail bodies to cross.

Onions: Medium ★★☆

Spanish red onions, small white onions and large brown onions can all be grown at home. Many of the varieties have unique timing for planting and harvesting, making it possible to plant onions most of the year in most climates.

Plant seeds directly into rows and thin out smaller seedlings. Alternatively, plant seeds into seedling trays and transplant into your veggie bed when plants are about 5cm tall. Onions like rich soil, full sun and regular watering – not water logging.

Harvest onions when the tops start to die back. Don't leave onions in the ground as they're prone to rotting. Cook your onions by roasting them in wedges, alongside other veggies and meats. Use home grown onions when you're making zucchini or cucumber pickle.

Silverbeet and Rainbow chard: Easy ★☆☆

Looking for something that's super easy to grow? Here's your new thing. Silverbeet and Rainbow chard are very forgiving and will grow in almost all conditions. They likes rich soil, full sun and regular water — sounds like most vegetables? Think again. Silverbeet and Rainbow chard are surprisingly tolerant of a wide range of conditions. They will live and produce lovely leaves in average soil with part-shade and partly neglectful watering. Watch out for snails and slugs when the plants are young.

Plant seeds or seedlings in early autumn and your plant will produce for a few years. Plants don't go to seed in a way that stops them from producing leaves over the summer.

Grow at least six plants so that you have enough to make vegetarian cannelloni or Palak paneer from autumn through to spring.

WINTER IN THE VEGGIE PATCH

Winter's midday sun is calling. Time to get into the veggie patch and grow happy as the veggies grow tall. There's a different pace in the winter veggie patch. It's gentle. It's slow. It's full of the colour green.

Broad beans: Easy ★☆☆

These super-hardy, nitrogen-fixing beans are a real joy to grow. Plant seeds directly into your veggie patch anytime from late autumn through winter and into early spring, spacing plants in two rows, about 15 cm apart. Broad beans prefer full sun and lightly manured soil, but they're not fussy and will grow in almost any conditions if it's cold.

As they grow, broad beans can benefit from being loosely tied up with string around the outside of your double row. Broad beans have a lovely pea-shaped flower that grows from the bottom of the stem. Flowers then transform into long pods, which contain the broad beans. Harvest your broad beans early, while the pods contain tender, sweet beans. These early broad beans can be eaten raw or lightly steamed. If you leave the pods on your plants then your broad beans will grow long and fat. You'll need to pod these beans and remove the outer part of these older beans.

Garlic: Easy ★☆☆

Home grown garlic is crisp, strong and can be plaited elegantly for storing at home. It's super-easy to grow and, as a bonus, growing your own garlic is a money saver.

Garlic likes very rich, well-draining soil. Mix manure as well as compost and worm casings to prepare the soil. Garlic bulbs grow from individual cloves, so you simply need to source locally grown bulbs from your grocer, the farmer's market or a garden centre.

Plant individual cloves about 5cm deep in the soil and about 10cm apart from each other. The bottom of the bulb side should be facing down. Garlic grows slowly, so good mulch and regular weeding are essential for success.

Garlic bulbs need low temperatures to mature, so they're not suitable for warm climates. Plant bulbs in late autumn or early winter. Harvest when the tops turn brown, in the springtime.

After you've pulled up the bulbs, let them dry off for a day in the sun, then it's time to chop off the roots, clean off the outer layer of garlic 'skin' and dirt and plait your garlic. Plaited garlic in your kitchen. So French. So chic.

Leafy greens: Easy-Medium ★☆☆-★★☆

Lettuce, English spinach and Silverbeet are delicious leafy greens that thrive in a late-winter veggie patch. For the best results, plant seedlings in August for a crop in the springtime. If you're growing from seed, raise your plants indoors until they're well established.

Leafy greens like soil that's enriched with plenty of compost, a sunny location and regular watering.

Vertical Garden

Want to surround yourself with green? Transform a drab wall on your balcony or entrance way into a vertical garden. With inspiration from the world's hipster café scene, vertical gardens are the perfect way to garden in high density homes. The only natural ingredient that you need is sunlight. The best walls for vertical gardens are north facing, with plenty of sunlight. However, with the right plants and nurturing, vertical gardens can thrive on any wall. All vertical gardens require a bit of love. Put them somewhere that you'll remember to water.

After finding a wall, work out how you're going to attach some pots. You can buy ready-made vertical garden frames from most garden outlets. Or... be creative. Use wire and some nails to attach pots that aren't currently in use. Consider putting your vertical garden on a rack that can be moved, as the seasons change. This will let you avoid the extreme heat in summer or frosts in winter.

Fill your pots with beautiful soil, rich and well-draining. Now get planting. The classic vertical garden is herbs, like parsley, mint, coriander and rocket. Branch out from the classic with some flowers or strawberries.

Flowers in the vegetable patch

Cut flowers bring sunshine, love and whimsey into a home. Why not grow your own? Consider growing a row of flowers, beside each row of vegetables. It will look gorgeous, improve pollination and supply you with cut flowers. The best cut flowers for growing between vegetables are annuals, grown by seed. Cosmos (white or mixed colours), poppies, cottage garden mix, sunflowers and marigolds are all fabulous. Flowers generally like similar conditions to vegetables. Most flowers prefer full sun, rich soil and regular watering.

KIDS IN THE VEGGIE PATCH

Snap off a snow pea, nibble on some kale and squeal in delight at butterflies. There's so much for kids to love in a veggie patch… including squirting water from the hose everywhere and making mud pies. The joy of planting, growing and harvesting can be experienced at every age. Here's how to get started on a kid's veggie patch.

Step 1. Location, location, location

Most importantly, the veggie patch needs to be in a place where you and your kids want to hang out. Spring and summer veggies will like a spot that gets plenty of sunlight but is sheltered from the hot afternoon sun. Too much

shade and your veggies will have their growth stunted. Too much sun and they'll bake on a hot summer day.

Step 2. Create a garden bed full of rich soil

Veggies like to eat lots and lots of organic matter. Most are happiest growing directly out of compost. Consider adding about 20cm of organic matter (compost or veggie soil) across the top of your regular soil, then turning it with a fork to lightly combine the existing soil with your layer of organic matter. Alternatively, create a raised bed with straw bales around the side and fill it with rich soil. Little veggie gardeners can turn the soil with a fork, just make sure they're wearing gumboots. They might also like to help by adding a little water to the veggie patch… and making a mud pie, or two, along the way.

Step 3. Get planting

Ask your kids what they'd like to grow and eat. Then consider veggies that will reward with little work. Here are my top picks in a spring or summer kid's veggie patch.

- Radishes. Radishes are fantastic because they're quick to grow. Plant seeds and just 4–6 weeks later, you'll be harvesting gorgeous red or white gems from underneath the soil. Kids love harvesting radishes. Some kids also love eating them.
- Rainbow chard. Great colour and the perfect ingredient for cheese and spinach triangles. Plant seeds and thin them out. Rainbow chard is one of the hardiest veggies, producing heaps of leaves right through until autumn.
- Peas and beans. So delicious when they're freshly harvested. Plant seeds directly into the soil and pop up a climbing frame. Most varieties can grow to the height of a person, if your frame is high enough. Sugar snap or snow peas love mild spring times and don't mind the frost. Beans, particularly blue lake climbing beans, are perfect in the summertime.
- Perennial herbs from seed. Fantastic for exploring new flavours and many produce lovely flowers. Rocket is fast growing and has a lovely white flower. Italian and curly leaved parsley can be grown side by side for flavour comparison. Dill has a delicate flavour and lovely, delicate flowers in the summertime.

- Flowers. A spectacular addition to every veggie patch and a great way to attract pollinating insects to your garden. Sunflowers need quite a bit of space per plant (think height) and take a few months to reach full maturity. Kids will love being able to look up towards a huge sunflower head that came from a seed they planted in the ground. Marigolds, when planted from seed, can form a thick border around the outside of your veggie patch. They're also super easy to grow and self-seed if you let them reach full maturity.
- Cherry tomatoes. It's so much fun to go hunting for ripe cherry tomatoes amongst the tangle of tomato vines. Choose any variety of small tomatoes including *yellow grape* or *tiny toms*. For best results, grow your tomatoes next to garden steaks and tie limbs to the steaks, so that you lift the vines off the ground.

Share the joy of being outside and in the veggie patch with kids… and just maybe they'll discover a new vegetable that makes it onto the family menu.

GROW YOUR OWN FRUIT IN EVERY SEASON

Home grown fruit tastes better. Juicy, crunchy, sun kissed... Yum. Yes, there is a reason. It's about the way your fruit ripens. Fruit ripens for longer on your trees at home, creating a sweeter and more complex flavour. Commercially grown fruit is picked early, so that it can ripen as it's transported to cold storage, warehouses and the supermarket. Eventually, it'll make it to the fruit bowl in your kitchen.

The pleasures of home-grown fruit are so much greater than just eating. It's the joy of being showered in apple blossoms, of watching your fruit grow and the reward of careful guarding from hopeful birds. With a little planning, you can enjoy home grown fruit throughout the year.

Citrus

Use your patio to grow citrus in a pot. Lemons can produce fruit throughout the winter months, from late autumn to early spring. Extra cold temperatures bring out a new sweetness. If there's frost where you live, keep young citrus in a pot and close to your home. After a few seasons, they'll be a bit bigger, acclimatised and ready for the garden.

Lemons are great to have in excess because they're so versatile. Use them when flavouring your evening meal. When there's extra, it's time for lemon curd, lemon tart or lemon meringue pie. Limes are the exotic, tropical lemon flavour. Grow Tahitian limes for their fruit, or Kaffir limes for their leaves, used in Asian-style cooking.

Consider growing oranges and mandarins. Keep them (very) well-watered throughout the fruit's growing period to create juicy fruit.

If you love making marmalade, cumquats are for you. They're easy to grow and will crop in almost any conditions.

Some nurseries offer citrus trees that are grafted with oranges and lemons on the same tree, a fruit salad tree. Fabulous for trying new things and for small spaces.

Mulberry

Every child needs a mulberry tree. They're big trees and great for kids to climb. Black mulberries are also great for kids to eat, if you don't mind a bit of mess. Mulberries produce sweet, black berries in the springtime. They're frost sensitive, so a late frost can reduce your crop significantly. Mulberries produce fruit in late spring and early summer.

Classic orchard fruit

Apples, pears, quinces, plums and nectarines are all classic orchard fruit. They all need a period of cooler weather in the wintertime and are suitable for cool and temperate climates, not the tropics. In the cooler months, classic orchard fruit trees are dormant. They lose their leaves and go to 'sleep'. It's in this dormant phase that you plant them out as bare root saplings.

Classic orchard trees can produce fruit from mid-summer, through to late autumn.

Feijoas

This sweet, green fruit is sometimes known as a pineapple guava. Peel away the tough, green outer layer and enjoy the sweet seeds and flesh. Feijoas are a hardy, evergreen bush. They're commonly used as a hedging plant. Without water, they produce fruit that's about an inch long, thin and with a small edible part inside. If you put your grey water onto a tree in March, as the fruit are growing, you can produce fruit the size of a cricket ball. Feijoas fruit from mid-autumn to early winter.

Figs

Figs have just the right amount of sweet, are packed full of flavour and have an interesting texture. Net your fig tree to protect it from birds and children. Keep the fig tree small enough to net by giving a good, hard prune each year. Figs produce a first flush in late summer then the bulk of their fruit in autumn.

Grapevine

Grapes have been cultivated for eating and wine making in the Mediterranean for more than 5,000 years. There the summers are hot and winter is cold and wet. The hardest part of placing a grape is finding a part of the garden for your grapes to stretch out. You can trellis them along a fence or along the side of your deck. If you want to create a garden feature with your grapes, grow them over a pergola so that the bunches drop down in a glorious early summer display. Most grapes like full sun and rich, well-drained soil. They have deep roots, so dig deep when preparing the soil. Harvest your grapes from summer through to autumn, depending on the variety.

Pruning fruit trees

Good pruning is essential for healthy, productive fruit trees. Pruning creates spaces inside the tree for birds and larger insects to eat smaller insects that could harm your tree. It also creates airflow, preventing mildew related diseases. Pruning also improves fruit production and can make it easier to harvest fruit. To keep your tree healthy, prune on a day when it isn't likely to rain. Three steps to excellent pruning:

1. Remove dead wood.
2. Reduce the tree's size. Remove any new leader branches (branches that are pointing directly up) and take about one third of the length off the branches, or enough so that the fruit is easy to harvest.
3. Shape the tree. Create space in the middle of the tree for air to flow and birds to hunt for insects.

CREATE AN EDIBLE OASIS

It's a shady summer afternoon underneath the grapevine. There are bunches of sweet grapes ready for harvest, large terracotta pots with flourishing fruit trees and a few small planter boxes bursting with basil, parsley and sage. It's an edible oasis! Here's how you can create your own on a patio or deck.

Step 1. Get started with an edible vine

Choose from grapes, kiwi fruit or passionfruit, depending on what you like to eat.

Step 2. Add some fruit trees in a pot

Citrus trees grow nicely in pots. They also look beautiful with their dark green leaves and brightly coloured fruit. Choose large terracotta pots, rather than wine barrels, as the terracotta holds moisture better — saving on both water and watering time. Citrus trees are frost sensitive, so put citrus pots against your home in the wintertime and in the dappled sunshine for the summer. Consider a range of different citrus trees, including lemons, Tahitian limes, kaffir limes, oranges, mandarins and cumquats (for making jam).

Step 3. Create space for herbs

A box that's bursting full of luscious, leafy herbs gives you a reason to harvest from your edible oasis all year round. Have two boxes, one that's full of perennials — herbs that die back in winter and re-grow in summer — and one that you plant for each season. In the perennial box, consider sage, chives, oregano, thyme, French tarragon, winter mint and Vietnamese mint. Most of these perennials are best grown from cuttings or small plants. In the box for annuals, consider planting seeds for summer with basil and parsley. In the wintertime, consider planting coriander, dill and parsley seeds. Don't be shy when you harvest your herbs. Most herbs can be cut right back, to just 4–5 leaves, and they'll bounce back. Don't be afraid to let excess plantings flower, most herbs have pretty, delicate flowers.

Step 4. Feed your oasis from home

All the plants in your oasis will appreciate a rich soil or liquid fertiliser applied in autumn and spring. The easiest way to create this soil or fertiliser is to make your own using a worm farm or Bokashi bucket. Both take food scraps from your kitchen and neither smells when its properly working.

Step 5. Make it beautiful

A string of fairy lights, or Moroccan-style tea light candles, can create magic. Add a comfortable outdoor setting beneath the vines and amongst the fruit trees. Your setting doesn't need to be new, consider sourcing something rustic that's second hand. Simply add a lick of paint or a tablecloth to complete your beautiful, edible oasis.

SAVE YOUR GARDEN FROM COMMON PESTS

Luscious green leaves and sweet, juicy fruit make the efforts of gardening worthwhile. This bountiful harvest is just what your garden pests were hoping to eat. Months of preparation, planting and growing can quickly disappear if you don't protect your garden from pests. Don't be disheartened. Try these five ways to protect your garden from common pests.

1. Prevention is the best cure

Strong, healthy plants are less likely to fall victim to many pests and diseases. Locate your garden in full sun with a lovely rich soil and rotate your crops each season. Look after your plants with a regular watering and a fortnightly dose of plant tonic. Interplant your vegetables and fruit trees with flowers. They look gorgeous and attract the predators of many pests. Finally, give your plants enough space for air to flow, and for predators to access any pest outbreaks. This is just as important in your vegetable patch as it is for fruit trees.

2. Slugs and snails

They may look small and innocent, but these critters can eat through a whole row of seedlings or half a lettuce in just one evening. The most effective organic treatment is to set traps. Slugs love beer so cut the bottom 4cm off an empty plastic bottle and half fill it with beer. Slugs can get into the beer, but not out. Snails love to hide under a moist brick, so pop a brick near your garden and cover it with a piece of cloth, then check it daily for snails. Feed any slugs or snails that you trap to chickens. Crushed

Pear and cherry slug damage

Snails hiding on the outside of a veggie patch in the shade of overgrown comfrey.

eggshells around the base of plants are a fabled slug and snail barrier. They work fine when it's dry but are ineffective after rain or heavy dew. While it's not technically organic, snail and slug pellets that contain an iron compound are available from most garden stores. They break down into non-poisonous by-products. Simply sprinkle the pellets around the plants that you need to protect.

3. Sap sucking aphids

These tiny bugs will settle in their hundreds on tender new stems or new leaves. They suck the sap and exude a sticky substance, on which black mould sometimes grows. Roses are a favourite for aphids and new buds can be covered (and destroyed) by these little critters. A strong jet of water will dislodge and kill aphids. Try a home-made white oil spray by mixing one cup of sunflower oil with half a cup of dishwashing liquid. This is your concentrate. Place a tablespoon of the concentrate into a litre spray bottle that's filled with water and spray the aphids. White oil suffocates the aphids. You can also buy a range of commercial, organic approved sprays in your local garden store.

4. Cabbage moth/butterfly

Pretty white butterflies in your vegetable patch are a bad sign. They're looking for a brassica plant, like cabbage, broccoli, kale or cauliflower; the perfect location for their eggs. Cabbage moth caterpillars eat voraciously and make holes and take chunks out of your luscious brassica leaves. A single caterpillar can eat half a leaf in just one day. The good news is that effective, organic treatment is available. There's a biological control, a bacterium, that you mix with water and spray onto leaves.

5. Wildlife and birds

Fruit trees with a plentiful harvest can be shared with birds and other wildlife, if you're feeling generous. If you want the full harvest, physical barriers are the most effective way to prevent birds and wildlife. Smaller trees and gardens can be netted — quite a bit of work at the season's beginning. You can also try preventing wildlife from entering your garden in the first place but knowing their route and cutting it off by adding a fence, trimming branches or placing corrugated iron around the trunks of trees or posts that they climb down. Some wildlife are repelled by strong smells and tastes in the same way that humans like and dislike different smells. For example, in Australia, you can buy commercial possum repelling sprays in most garden stores or try making your own from garlic spray. Finely chop four cloves of garlic and place them in a litre of boiling water. Let the mixture stand overnight then spray it onto the possum's entry route to your garden or directly onto affected leaves.

CULTIVATE PLANTS FROM CUTTINGS

Plants that flourish from cuttings are almost indistinguishable from magic. Rambling jasmine flowers, gorgeous geraniums, rosemary bushes and curry plants can all be grown from cuttings. Cuttings create a plant that is a genetically identical copy of the 'mother' plant. This means it will be the exact colour and smell of a favourite flower.

The easiest plants to grow from cuttings are small shrubs, like rosemary, sage or geraniums. Late spring is the perfect time of the year for softwood cuttings.

Here's how to do it:

Step 1. Take the cutting

Using clean garden equipment, remove a softwood section of the plant that you wish to cultivate. Aim for a length of cutting that includes at least 10 leaf nodules and is a little firm. You'll need more than just the 'sappy' growth at the end of a stem.

Step 2. Prepare for cultivation

Gently remove leaves from the bottom 5–8 nodules. You will now have 3–4 leaves on top and a bare stem below. You don't want too many leaves on top as they can dehydrate the cutting. If you're super keen to have a high strike rate, now is the time to dip the bottom of your cutting into some plant hormone to stimulate growth. Using hormone is optional, it just improves the likelihood of your cutting surviving.

Step 3. Plant into a pot

Place your cutting in a pot containing rich, well-drained soil. Water it in with some liquid seaweed tonic or liquid worm fertiliser. You'll need to tenderly water your cutting over the next 5–8 weeks while it creates roots and becomes a new plant.

Step 4. Plant in the garden

After your cutting has developed its own roots, you'll notice that it starts to grow new leaves as well. Now is the time to re-pot or plant it directly into a garden bed.

For those who are adventurous... or lazy gardeners, some plant cuttings can be placed directly into a garden bed. This lets you skip step 3, planting into a pot. Geraniums have incredible resilience and almost every cutting will create a new plant when placed directly in their new location. Just be sure to water them well in the initial few weeks, including with a dose or two of liquid seaweed tonic or liquid worm fertiliser.

CULTIVATE PLANTS FROM CLUMPS, PUPS AND RUNNERS

Lush bushes of mint, spectacular sunflowers produced by Jerusalem artichokes and delicious red strawberries can all be cultivated from plants in your own backyard… or from the backyard of neighbours and friends. There's no need to search nurseries for that special variety, simply take a cutting or separate the roots then put your new plant directly into the soil. Late spring or early summer is the perfect time of year to get started.

Here are four simple approaches to cultivating plants from clumps, pups and runners in your own backyard:

1. Separate a clump of herbs

Mint, oregano, thyme, tarragon and comfrey are all herbs that grow in clumps. You'll notice that a few stems emerging from the earth quickly double or triple to create a herb clump. To separate your herb clump, just dig out a portion of the clump with its roots. Leave at least six stems behind to maintain your existing plant. If you want to create lots of plants, you can separate the stems that you've dug up and plant them individually. If you're lazy, just plop the clump into a pot or directly into your new herb bed. Water them and give a generous dose of liquid seaweed tonic or liquid worm fertiliser. This will help both your existing plant and the new plants to settle successfully.

2. Separate clumps of large edibles

Raspberries and blackberries grow on canes in clumps. Every spring you'll see small plants called suckers emerge from the soil adjacent to your raspberry or blackberry cane. Transplant these small plants in spring when they're a few feet tall and have well-developed roots.

Jerusalem artichokes are grown for their potato-like tuber that tastes slightly sweet, slightly nutty, and create awkward moments with foul smelling flatulence! They also have lovely sunflowers. Jerusalem artichokes are extremely hardy plants. They like full sun but will grow in almost any conditions. Separate tubers in early autumn, after flowering. The following spring, your tubers will emerge from the soil to create a lovely tall sunflower… and more tubers for you to eat!

3. Transplant succulent pups

Super easy doesn't begin to describe what it is like to look after succulents. Ignoring them is the surest way for them to thrive! Succulents have thick and fleshy leaves that are adapted for storing water. Cacti are one of several families in the succulent group. They're great plants to have both indoors and outside. Succulents prefer full sun, or bright indoor light and well-drained soil. Overwatering is the most common cause of indoor succulents not thriving.

Succulent pups are baby plants that grow up from the base of their 'mother' plant. They're usually mature enough to transplant after 2–3 weeks once they've developed roots. Delicately remove succulent pups and replant them in the pot or garden bed. Water them in with liquid seaweed fertiliser or homemade nutrient tonic.

4. Create new strawberry plants from runners

Runners are long stems that grow parallel to the ground with the ability to put down roots. In late spring and early summer, you will notice strawberry plants putting out runners that turn into new mini plants. After the mini plant has put down roots, you can cut the runner and transplant it into a pot or garden bed. Strawberry plants put out more than one runner per plant, so if you've got a keen eye, you can more than double the number of strawberries in your patch every year! Juicy, red strawberries are simply delicious in every way. Perfect in fruit salad, punch or to eat straight from the garden.

WATER EFFICIENT GARDENING

Water is an essential ingredient to every garden. It transforms drooping veggies into a lush and crisp harvest. It's the key for a bumper crop from backyard fruit trees. The ancient proverb, water is life, is just as true today as it has been for centuries.

Here's how to be smart with the water in your garden.

1. Design for your climate

What do you really need from your garden? Is it veggies and fruit trees, privacy screening, a place to play cricket or a wildlife habitat? Seek advice from local nurseries about plants that suit your climate and meet your needs. Be ruthless with the size of your lawn. A manicured lawn is a water guzzler — the enormous four-wheel drive equivalent in the garden. Be sparing with your fruit trees. The best crops for many trees, especially stone fruit and citrus, require lots of water in the middle of summer as the fruit is developing.

2. Source your water for (nearly) free

Install a rainwater tank — the largest that fits in your available space. Rainwater is a perfect water source for your veggie garden. If you're using a drip irrigation system, check the pressure in your drippers, all the way to the end of the line. You may need several shorter drip lines instead of one long line. You may also want to connect your drip lines to mains water, as a back-up for times when the tank is empty.

Recycle your water onsite. Water that's come from your shower, bath and washing machine is also known as 'grey' water. It's been used, but it's still suitable for a second on-site use, as long as it's not sitting around for more than a day. Brand new, environmentally designed buildings often have grey water diverted to an onsite tank that pumps out to a garden every 24 hours.

Grey water is perfect for fruit trees, wildlife habitat and lawns. The simplest grey water system starts with a switch in your laundry that lets you choose the drain or a pipe out to your garden. The pipe to your garden can either go directly onto your garden, or into a large tank on wheels (think modified wheelie bin) that you move to the location you'd like to water. Some people improve the quality of their grey water with a natural-style reed bed, frog pond and mini wildlife haven.

3. Use your water well

Deliver the precious water that you have, directly to the plants that need it. Try an above-ground wicking bed for your veggies. They take quite a bit of construction, as the bed has layers of large stones, gravel, soil and compost. Once built, they're a fabulous way of delivering the perfect amount of water to

Drip irrigation pipes.

your veggies, as the water is stored between the rocks and is accessible to your veggies 24/7. You simply top up the reservoir beneath your veggies.

If you prefer a traditional veggie bed, irrigate using a dripper that's just on the surface of the soil. Drip irrigation is a very efficient way of delivering water to the soil. Space your plantings to match up with each drip hole.

Hand watering is another efficient way to deliver water directly where it's needed. It's also a beautiful way to connect with your garden. Nip those weeds before they grow tall. Watch your harvest start as a flower and transform into a crisp snow pea. The hardest part of hand watering is finding time.

4. Mulch

For every tree, veggie, habitat or screen planting… be generous with mulch. Mountains of mulch are fabulous in every garden as they prevent water that's in your soil from evaporating. Mound your mulch up high, at least 10cm. In addition to keeping your plants hydrated, mulch breaks down into food for your plants. I guess it's like an edible plant moisturiser!

CHICKENS

Backyard chickens are the ultimate domestic farm animal for large urban backyards. They make a delightful clucking sound. The eggs are superb. They turn scrap food into compost. They even give you love… or maybe they run towards you because you have their food.

Here are the basics for getting started with backyard chickens.

1. A secure place for chickens to sleep and lay eggs

Pick up a second hand hen house, build your own or buy something new from a pet shop. Some people like to get creative with their hen houses. Consider whether your hen house will move around the garden each week or with a summer and winter location. Alternatively, it might have a permanent location. Keep your chickens locked up each night as foxes and feral cats roam throughout the suburbs.

Inside the hen house, you'll need space for roosting and space to lay eggs. Keeping these two spaces separate reduces the likelihood of poop on fresh eggs. Chickens prefer to roost on the highest possible point, so make sure their roost is higher than the laying box and they'll get the idea. Cover the space underneath their roost with old newspaper and change it weekly, to keep the chooks healthy. This newspaper covered in chicken poo is great in most composts.

2. A source of water

Pet shops or farm supply stores can help here as they sell special poultry water feeders. Chickens love fresh, running water. Change the water in a feeder each week to keep your chickens healthy.

3. Food

You can get away with feeding your chooks by hand every day if you're always around. Chooks will survive on family kitchen scraps and foraged bugs and leaves.

If you want lots of eggs, consider supplementing your chicken's diet with grain or an especially created chicken 'scratch mix'. Start low-tech by just throw them a handful or two each morning. If you like to take a few days off hand feeding, invest in a chicken feeder that is chicken friendly and rat/mice proof.

4. A fenced yard

Will you be letting your chickens out into the yard or keeping them enclosed? To be considered commercially free-range, each chicken needs upwards from 4 m^2 per bird. Chickens love to eat and dig up the veggie patch. Perfect after you've finished the harvest, not before. Consider fencing your chickens into the orchard area, fencing your veggie patch or closely supervising chickens when they're near your veggie patch.

5. Choosing your chickens

When it comes to egg laying, Isa Brown chickens are the most productive in their first 2–3 years. Their eggs are light brown. Other breeds lay less eggs per week, for a slightly longer period. Leghorns and Australorps are known for being solid egg layers, over their 3–4-year life. They lay a variety of white, light brown and dark brown eggs.

If fancy eggs are your thing, rather than productive chooks, investigate breeds that lay eggs that are blue or green like Araucanas and Ameraucanas.

Source your chickens as fertilised eggs (with an incubator), day old chicks, pullets (15 weeks old, females only) or at the point of lay (18–20 weeks old, females only). The smaller chickens may include roosters, so you'll need to decide on their future when their sex becomes clear at around the 18–20-week age. If you've already got chickens, don't introduce new chickens that are younger than 18 weeks old as the stress of sorting out the pecking order could result in death.

THE BUZZ ON BACKYARD BEES

Heard the buzz about bees? Well, it's created by their wings beating 190 times per second… they communicate with each other by dance and are responsible for pollinating at least 70% of the world's food. As a bonus, they produce sweet, delicious honey.

Thinking about keeping backyard bees? Here's how to get started.

1. Choose your bee type

European honeybees, *Apis mellifera*, are excellent honey producers. They're used in most of the world's managed hives. In a backyard setting, they'll produce an average of 75kg of honey per hive, every year.

Not so sure about the sting? In Australia, there are many, many species of native bees. They're smaller than European bees and all are stingless. The downside is in honey production, it's significantly less than European bees. Some species produce as little as 1kg per hive in a year. Native bees are only suitable for tropical and sub-tropical areas, about as far south as Sydney. Anywhere else is too cold.

2. Choose your hive type and equipment

Traditional European beehives consist of a box with removable frames. To harvest the honey, you'll need to remove each frame. This involves suiting up and smoking the hive.

There's a new type of hive on the market for European bees, the *flow hive.* It was invented by an Australian beekeeping family and is designed to minimise the disruption to bees which occurs in honey harvesting. It's pricier than a regular hive, but advocates say that it creates a more relaxed environment for your bees.

Native beehives look like those created for European bees but are typically smaller. You won't need a bee keeping suit or a smoker to harvest the honey, the bonus of keeping with stingless bees.

3. Get your garden ready

In the morning, your bees will emerge from their hive and take off towards the sun. So, choose a place in your garden that's got a nice, wide, north-east facing flight path. If possible, shelter your hive from the extremes of cold and heat by placing the hive underneath a tree and in a place that isn't easily disturbed.

Got children? Then you might want to make sure that the hive is behind a fence or in a part of the garden that they and their friends are unlikely to disturb. A small boy with a handful of stones is a recipe for unhappy bees and (possibly) a very sore boy.

Bees will need a permanent water source so they can walk down to the water's edge and drink. A bucket doesn't work as the sides are steep. Find a shallow dish, fill it with water and angle a plank of wood into the water, creating a ramp for the bees to access water.

4. Do a beekeeping course and register

Most local beekeeping clubs run introductory courses that are half a day long, with longer courses for those with a little more experience. Connect with your local beekeeping association to find out more.

NATIVE BACKYARD BIRDS

Are you awake in the early morning, when the world is filled with a chorus of bird song? There's the musical song of magpies, a screeching of parrots and sweet tweeting of wrens that welcome each new day, if you're awake in time.

Here's how you can encourage backyard birds to enhance that beautiful morning chorus.

Plant a habitat garden

A neatly manicured lawn is nice for lawn bowls or tennis, but it doesn't encourage birds. For more birds in your garden, start with some dense, native shrubs that provide shelter for smaller birds as well as nectar from their flowers. If you've got space, try a medium-sized tree habitat and a source of food.

A nesting box to encourage backyard birds.

It's a bit unconventional, but if you have space, consider large fruit trees as a way of attracting large birds. When an apple tree is laden with fruit, you can harvest from the bottom of the tree, leaving the treetops to the birds. The flash of colour and raucous chatter can be a welcome addition to your backyard.

Install a nesting box (or two)

One of the biggest threats to native bird populations is land clearing and habitat loss. When land is cleared and established trees are removed, the hollows for nesting cannot be replaced by young trees. Enter nesting boxes. To encourage native birds to nest, attach a box to one of your taller backyard trees. Different box sizes will fit different birds, so choose your box carefully.

Best pets

Even the nicest cats are bad news for backyard birds. Indoor-only cats are the latest in pet trends and native birds love it. Unfortunately, there are still domestic and feral cats roaming our urban areas day and night.

KITCHEN GARDEN DELICIOUS

AUTUMN KITCHEN GARDEN

Ripe figs, luscious leafy greens and still more zucchinis… It must be autumn in the kitchen garden. A great season for growing. The scorching heat of a mid-summer's day has past, leaving just a lovely warm soil and mild days — perfect growing weather. Tender, crisp leafy plants love autumn.

In the kitchen, it's time to put the last of summer's garden gifts into a compote, a cake and a pickle.

Rhubarb and lemon compote

500g rhubarb stalks
1 large lemon (peeled and juiced)
2 bay leaves
5 cloves
1 cup brown sugar

Chop the rhubarb stalks into 2.5cm lengths then put them into a medium sized cooking pot. Add the juice and peel from the lemon, bay leaves, cloves and brown sugar. Bring to the boil and simmer for about 5 minutes until the rhubarb is tender. Remove the cloves then bottle in sterilised jars or freeze in batches.

Enjoy with thick yoghurt, cream or ice cream.

Rhubarb cake

300g rhubarb stalks
2 tbsp brown sugar
¼ tsp ground cloves
2 free range eggs
¼ cup milk
¼ cup melted butter
⅔ cup caster sugar
1½ cups plain flour
3 tsp baking powder

Prepare the rhubarb by chopping stalks into 2.5cm lengths then place them into a medium sized cooking pot. Add the brown sugar, cloves and a splash of water to prevent sticking. Stew on a low heat until most of the rhubarb is soft. The rhubarb should be soft but not swimming in liquid. If there's liquid, gently drain it off.

In a separate bowl combine the eggs, milk, melted butter and caster sugar. Mix well. Add the flour and baking powder then mix into a smooth batter.

Heat oven to 180°C. Grease a 20cm springform, round cake tin.

Pour half of the batter into the bottom of the cake tin. Place rhubarb mixture on top, reserving 6–8 pieces of soft rhubarb. Dollop remaining batter onto the top of the rhubarb mixture, to create a second layer in the cake. Top with reserved rhubarb pieces.

Bake for approximately 20 minutes, until both layers are cooked.

Place strawberries and seasonal, edible flowers on top, then top with icing sugar.

Fig and almond upside-down cake

½ cup wholemeal flour
½ cup white flour
½ cup almond meal
1 tsp baking powder
1 cup white sugar
3 eggs
½ cup yoghurt
½ cup canola oil
½ tsp vanilla essence

Combine the dry ingredients in a large mixing bowl. Combine the wet ingredients in a separate bowl by whisking together the eggs, yoghurt, canola oil and vanilla essence. Gently pour the wet ingredients into the dry ingredients and combine.

Now for the upside-down part of this cake. Line a 20cm round tin with baking paper and grease the bottom with butter. Thinly slice three large figs and arrange the slices elegantly on top of the greased baking tin. Sprinkle with a tablespoon of caster sugar.

Gently pour the cake mixture on top of the figs. Bake in a moderate oven (160–180°C) for about 40 minutes, until a skewer comes out clean. Turn the cake upside down and serve with Greek yoghurt or cream.

This recipe works well with any autumn fruit, including apples and plums.

Pumpkin and barley salad

Serves 4 as a main meal and 8 as a side salad

¾ cup barley
1 tsp vegetable stock powder
1.5kg pumpkin
1 generous glug of olive oil
1 garlic clove (grated)
Salt and pepper for seasoning
6 cups of mixed green leaves from the garden (baby spinach and rocket work well together)
100g feta cheese

Dressing

½ cup extra virgin olive oil
1 tsp Dijon mustard
1 tbsp balsamic vinegar

Place the barley and stock powder in a saucepan along with 2 cups of water. Simmer for 35–45 minutes, until the barley is cooked and the water has been completely absorbed.

Dice the pumpkin into 2cm cubes and pop it onto a grill tray. Mix through the olive oil and garlic. Season lightly with salt and pepper. Grill on high for 15–20 minutes, until the tops are just a bit burnt and the pumpkin is mostly cooked. Let the tray of pumpkin sit in the grill for a further 5 minutes while all the pumpkin cooks through.

Now for the dressing. Mix the olive oil, Dijon mustard, balsamic vinegar and season to taste. Shake these up in a small jar so they're well combined.

Finally, put the salad together. Lay the mixed leaves from the garden on a platter, gently place the grilled pumpkin and barley on top. Crumble your feta cheese over the salad. Drizzle the dressing and gently fold the salad ingredients.

Yum! There will be enough pumpkin and barley salad to share between a few families at a BBQ or brunch.

Roast beetroot salad

Serves 4 as a main meal or 8 as a side salad

10 small beetroots
1 generous glug of extra virgin olive oil
2 cups cooked lentils
⅓ cup toasted sunflower seeds
4 cups baby spinach
Salt and pepper for seasoning

Dressing
Juice from 1 lemon
1 tsp. Dijon mustard
⅓ cup olive oil
¼ cup chopped tarragon

Scrub your beetroots then chop them into wedges and pop them into a roasting pan. This will probably fill a whole pan. Drizzle with olive oil and season with salt and black pepper. Roast at 180°C for about 40 minutes or until the beets are well cooked and the edges are starting to caramelise.

To make the dressing, combine all ingredients in a small jar. Season to taste then shake the jar so the ingredients are well combined.

Assemble the salad by placing the baby spinach on a serving platter, arranging the roast beetroot on top. Combine the lentils and toasted sunflower seeds then place them on top of the roast beetroot. Drizzle the dressing over just before serving and lightly toss the salad.

The lentils and sunflower seeds compliment the beetroot, giving this salad a real earthy flavour. Tarragon in the dressing gives the salad a lift and a zing.

Baby spinach, kale, broccoli leaves and rocket can often be interchanged in salad recipes. Use the combination of leafy greens available from your kitchen garden.

Pumpkin soup

Serves 6

1.5kg pumpkin
2 generous drizzles of extra virgin olive oil
1 onion
1 garlic clove
Celery leaves (optional)
4 cups warmed vegetable or chicken stock
400g cooked chickpeas
1 tsp Dijon mustard
1 tbsp honey
Salt and pepper for seasoning

Remove the pumpkin's skin and chop your pumpkin flesh into large chunks. Place your pumpkin chunks on an oven tray, drizzle with olive oil and season with salt and pepper. Roast your in the oven at 180°C for about 45 minutes. Your pumpkin will have caramelised on the bottom, creating a natural sweetness.

Meanwhile, finely chop the onion and garlic. Gently sauté with a drizzle of olive oil in a large saucepan. Add celery leaves, if you have them, in the final stages of the sauté process.

Add your roast pumpkin to the large saucepan, taking care to include the small, caramelised bits that like sticking to the pan. Add stock, chickpeas, Dijon mustard and honey. Using a stick blender, puree to create a smooth soup texture. Test for seasoning and sweetness, adjusting with salt, pepper, Dijon mustard and honey to your taste. Heat your soup to the point of simmering.

Serve with dollops of plain Greek yoghurt.

Chickpea and lentil salad

Serves 4–6 as a main meal

1½ cups cooked chickpeas
2½ cups cooked blue lentils (French lentils)
1½ cups herbs from your veggie patch
3 tbsp toasted pine nuts
¼ cup toasted seeds (pepitas, sunflower seeds and sesame seeds all work well)

Dressing

1½ tbsp white wine vinegar
¼ cup extra virgin olive oil
Salt and pepper to taste

Make the dressing by combining the white wine vinegar, extra virgin olive oil, salt and pepper in a small jar. Shake your jar to combine the dressing ingredients. Test to check that the seasoning is right for you.

Combine all remaining ingredients in a large salad bowl. Drizzle on the dressing, just moments before you're ready to eat.

Carrot and lentil salad

Serves 4–6 as a side salad

160g blue lentils (French lentils)
1 bay leaf
2 red onions
1½ tbsp extra virgin olive oil
1.2kg carrots
4 tbsp lemon juice
4 tbsp tahini
1 bunch of kale (about 12 stems)
2 tsp sesame seeds.
Salt and pepper for seasoning

In a pot, cover your lentils with about 2cm of water. Add a bay leaf. Bring to the boil and simmer for approximately 20 minutes, until the lentils are soft. Drain the lentils and you'll have approximately 2 cups of cooked lentils.

Slice the red onion. Sauté on a medium heat in about 1 tbsp of olive oil. Stir until the onions are translucent then set them aside.

Slice your carrots into thick sticks and sauté in about ½ tbsp of olive oil. Stir until the carrots are lightly cooked then set them aside.

Make the dressing by combining your lemon juice and tahini. Slowly add water to your lemon juice and tahini till the mixture is a runny consistency. Season to taste with salt and pepper.

Bring your salad together. Roughly chop your kale and place on a large salad plate. Place the lentils, sautéed onions and carrots on top then gently combine. Drizzle the dressing over the salad and top with sesame seeds. Delicious!

Quince paste

A picnic basket is not complete without cheese, crackers and quince paste. The sweet, aromatic quince flavour can transform an everyday cheese platter into something truly spectacular.

Quince paste is surprisingly simple to make. The key ingredient is patience as it needs to be simmered for about 2 hours on a very, very low heat, with occasional stirring.

If you really want to show off, make quince paste from your very own quince fruit. The trees are hardy cousins of apples and pears. Source your quince tree as rootstock from your local nursery in autumn or winter. Quinces will thrive in full sun and in well-drained, rich soil. They can grow just about anywhere, including in pots, and still produce fruit. For those who aren't yet feeling confident, you'll be pleased to learn that quinces don't need much water. While they do prefer water when establishing and producing fruit, they are also pretty good with neglectful watering. Enjoy beautiful, white blossoms in the springtime and in early autumn, simply harvest their bright yellow, knobbly fruit for your quince paste. When ripe, quinces have yellow skin, and very firm flesh.

1kg quinces (approx. 4 fruit)
500g white sugar

Step 1. Peel their skin, quarter and chop out the tough core then roughly chop up the flesh. Freshly chopped quinces have a truly delightful aroma reminiscent of peaches or mango and sweetness that is fit for royalty.

Step 2. Place the chopped quinces in a heavy based saucepan. Add 100ml of water. Stew the chopped quinces on a low heat until they are very soft. Stir semi-regularly to prevent the stewing fruit from sticking to the saucepan's bottom. This will take 20–30 minutes.

Step 3. Using a stick blender, puree the quinces in the saucepan.

Step 4. Add the sugar to the pureed quince paste and stir through. Place a simmer mat underneath your saucepan and return the saucepan to a very low heat for 2 hours. Stir occasionally. If you don't have a simmer mat, be a little more vigilant with your stirring and embrace the 'caramelisation' that comes with the quince mixture at the bottom of your saucepan. You'll know that your quince paste is ready to be removed from the saucepan when it becomes very thick and the colour has changed from yellow to a rich, dark red.

Step 5. Sterilise six small jars. Fill the jars with quince paste and set by cooling them in the fridge for two hours. Store in the fridge.

Enjoy your truly spectacular picnic experience, with cheese, crackers and the delightful flavours of homemade quince paste.

JAMS, PICKLES AND PRESERVES

Preserved lemons

5 large lemons
1 cup cooking salt
2 bay leaves
8 peppercorns
5 cardamon pods

Scrub the lemons then chop each into six wedges. Take care to remove any pips. Place the wedges into a large, non-reactive bowl and pour the salt on top. Using a wooden spoon, pound the lemon wedges, until most of the juice has been extracted. Lift out the lemon wedges and place them into a large, sterilised jar, along with the bay leaves, peppercorns and cardamom pods. Pour the salty lemon liquid into the jar to cover the lemon wedges and spices. Store in the fridge and use after six weeks, when the flavours have combined and the lemon rinds are soft.

Preserved lemon wedges can be finely diced and sprinkled on vegetables like cauliflower before it's roasted. They're also amazing with chicken or fish. Use the preserved lemon juices in a salad dressing.

Pickles

Crunchy, salty, sweet and tangy. Pickles are the ultimate burst of flavour that transforms an ordinary sandwich into something very special. They're also a great way to preserve a bumper vegetable harvest or to make the most of vegetables that are starting to wilt in the back of your fridge. Pickling has been used to preserve vegetables for thousands of years. In modern times, most home-made pickles are designed to be kept in the fridge, only some have enough vinegar and salt to create a classic preserve.

Cucumber pickle

800g cucumbers (any variety: Lebanese, long or apple all work well for this recipe)
1 small onion
1½ tsp salt

Pickling liquid
1 cup white vinegar
1 cup white sugar
1 tsp chopped dill
¼ tsp ground turmeric
½ tsp mild curry powder

Thinly slice the cucumbers and finely dice the onion. Place the cucumber and onion into an unreactive bowl. Separately, put the salt into a saucepan with ¼ cup of water. Stir, over a medium heat, until the salt is completely dissolved, then pour over the cucumber and onion mixture. Leave for 3 hours. Tip out the mixture into a colander and press with the back of a large spoon to remove as much liquid as possible.

Make the pickling liquid by combining all ingredients in a heavy based saucepan. Simmer over a medium heat until the sugar is dissolved. Pour the cucumber and onion mixture into the pickling liquid and simmer for a few minutes. Pack the pickle into jars, taking care that the liquid is on top of the cucumber and onion.

Bottle in sterilised jars and store in the fridge.

Cucumber pickle turns a cheese sandwich into something special. It's also a delicious accompaniment to roasts, served on the side.

Zucchini pickle

1–2 zucchinis (500g)
½ red onion
1 tbsp table salt
Sprigs of dill or tarragon
Bay leaves
Peppercorns

Pickling liquid
2 cups white vinegar
1 cup white sugar
1½ tsp curry powder

Thinly slice the zucchini using a vegetable peeler or mandolin. Created your zucchini slices in the shape that you want to eat — in sandwiches or salads. Long strips are great on sandwiches and round slices are perfect in salads. Finely slice the onion. Toss your zucchini and onion slices in salt. Leave in a colander for 1–2 hours while the veggies sweat out their moisture.

Create your pickling liquid by gently warming all the pickling liquid ingredients. Heat until bubbles start to form and the sugar is dissolved. Stand and allow to cool.

When your veggies have finished sweating, squeeze the zucchini and onions to reduce the moisture further. Pack your veggies tightly into sterilised jars along with a few sprigs of fresh herbs. Add one bay leaf and several peppercorns to each jar. Pour over the pickling liquid. It should completely cover the zucchini and onion mixture. Place your jars of pickled zucchini in the fridge.

Allow your pickles at least two days for the flavour to infuse. Enjoy the pizazz from homemade pickle in salads or on sandwiches.

Radish pickle

300g radishes

Pickling liquid
1 cup white vinegar
1 cup white sugar
½ cup water
1 tsp ground mustard
1 tsp whole coriander
2 bay leaves

Create your pickling liquid first by combining all ingredients in a saucepan. Heat gently to dissolve the sugar.

Finely slice the radishes then place into sterilised jars. Pour over the pickling liquid and place the jars in the fridge.

Allow your pickles at least two days for the flavour to infuse. Simple, no nonsense and delicious!

Garlic confit

4 garlic bulbs
2 tbsp white wine vinegar
3 basil sprigs or 2 bay leaves
Extra virgin olive oil

Carefully slice the bottom off each garlic bulb so that the bottom of each clove is exposed. Separate the bulbs, leaving the skin on. Place the cloves into a saucepan, add the vinegar, basil sprigs or bay leaves and then cover the cloves with olive oil.

Simmer this mixture gently for about 40 minutes, until the cloves are soft. Pour the garlic and oil mixture into sterilised jars and store in the fridge for up to 2 months. The low acidity in garlic makes it difficult to safely preserve, the addition of vinegar at about ⅓ of the garlic volume is essential.

Rub the garlic confit on bruschetta, stir through pasta or vegetables and stir through mayonnaise.

Green tomato chutney

1.3kg green tomatoes, chopped
2 brown onions, chopped
2 granny smith apples, chopped
2 cups white vinegar
1¼ cups brown sugar
1 tsp salt
2 tbsp mixed spice

Combine all ingredients in a heavy based saucepan. Bring to the boil and simmer for about 40 minutes. Stir every now and then so the mixture doesn't burn on the bottom and to let flavours combine. The chutney is ready when most of the liquid has evaporated.

Marmalade

Paddington bear is a fan of marmalade. When you make this marmalade, you might just find that there's a Paddington bear that comes out at night to devour it directly from the jar with their paws!

4 cups sliced lemons
4 cups sliced oranges
2 cups water
8 cups white sugar
75g pectin

This recipe needs 5–8 lemons and the same number of oranges. Use a mixture of blood oranges and navels to create a beautiful, rich coloured marmalade.

Wash and scrub the outside of your lemons and oranges. Slice your lemons and oranges in half, then into long quarters. Remove pips, then slice the quarters into eighths. These eighths are the perfect shape to be thinly sliced to create marmalade. Thinly slice the lemons and oranges.

Place sliced lemons and oranges in a large, heavy based saucepan. Add the water. Leave in the saucepan overnight to soften the lemon and orange skin.

Add the sugar to the soaked fruit and bring to a gentle boil for 15 minutes until the sugar is fully dissolved. Skim off any white 'scum'. Add the pectin and boil for a further 5–10 min.

Bottle in sterilised jars. Store in the pantry for up to two months… unless it's given away or eaten by Paddington bear!

Burnt Fig Jam

500g figs
500g white sugar
½ lemon (juiced)
Pectin

Roughly chopped the figs and place them into a heavy based saucepan along with the sugar and lemon juice. Bring the mixture to the boil, stirring to let the sugar dissolve and prevent the figs from sticking to the bottom. Simmer for 15–20 minutes.

Add pectin according to instructions on the packet for 500g of fruit. Check that the jam has set by seeing if a little glob on a metal spoon has a nice, jam-like consistency when given a moment to cool outside the pot.

Now for the burnt part. Let your fig jam cool completely. Then bring it to the boil again, stirring regularly, for 5–10 minutes to caramelize the sugar and 'burn' your jam. Watch the jam carefully in this stage as you're turning sugar into caramel and this mightn't be so good for your pot if you overcook.

How to sterilise bottles and jars at home

You can reuse glass jars and bottles by simply sterilising them in the microwave or on the stovetop at home.

Start by washing the jars and bottles well in hot, soapy water. Remove labels on the outside. You can do this by hand or in the dishwasher. The key to this stage is to remove all of the 'bits' of food that might cause contamination.

Sterilise in the microwave

Place a few centimetres of water in the bottom of a clean jar or bottle then gently rest the lid (yes, including metal lids) on top of the jar, leaving space for steam to escape. Microwave on high for 5 minutes then discard any water left in the bottom of the jar. Allow to air dry.

Sterilise on the stovetop

Place your jars, bottles and lids in a large saucepan, cover with water. Bring the water to boil for 4 minutes. Let the water and jars, bottles and lids cool down. Remove and allow the jars, bottles and lids to air dry.

WINTER KITCHEN GARDEN

Winter food warms our body and soul. Rich flavours energise, leafy greens help us to glow and hearty soups are perfect for batch cooking. Seasonal ingredients, like fresh broccoli, abundant citrus and pumpkins (stored from the autumn harvest) are the key to fabulous winter food.

Roast cauliflower salad

Serves 2 as a main meal for lunch, or 4 as a side salad

½ cauliflower
2 carrots
1½ tsp fennel seeds
1 cup mint leaves
1 cup cooked chickpeas

Dressing
Juice from 1 lemon
Extra virgin olive oil (¼ cup for the dressing and extra for roasting the vegetables)
Salt and pepper to season

Set your oven to 200°C. Chop the cauliflower into florets, leaving small florets whole and cutting larger florets in half. Chop the carrots into 1cm thick round slices.

Combine chopped cauliflower and carrots in a large baking tray. Sprinkle on the fennel seeds. Generously drizzle olive oil over the veggies. Season with a few grinds of salt and pepper. Evenly spread the cauliflower and carrots across the tray and roast for 30–40 minutes until the vegetables start to caramelise.

While the vegetables are roasting, make your dressing by combining the freshly squeezed lemon juice with ¼ cup of olive oil. Season with salt and pepper to taste.

After the vegetables are finished roasting, let them cool for 5 minutes. Combine the roast vegetables with the chickpeas and mint leaves in a salad bowl.

Immediately before serving, pour over your dressing.

Crunchy cabbage and jalapeno slaw

Serves 2 as a main meal or 4 as a side salad

2 cups roughly shredded red cabbage
1½ cups chopped celery
1½ cups coarsely grated carrot (about 1 large carrot)
1 cup torn rocket leaves
1 cup coriander leaves
⅓ cup flaked almonds
Extra virgin olive oil
Locally sourced jalapeno sauce

Place the red cabbage, celery, carrot, rocket and coriander leaves in a mixing bowl. Combine gently, using large spoons or your (recently washed) hands.

Arrange combined ingredients on a serving platter, then top with flaked almonds, a light drizzle of olive oil and a generous drizzle of jalapeno sauce.

Chargrilled broccoli salad

Serves 4–6 as a substantial side salad

1kg broccoli (2 large heads)
2 tbsp extra virgin olive oil
Salt and pepper for seasoning
¼ cup slithered almonds
1 cup chopped mint

Dressing
2 tbsp extra virgin olive oil
3 tbsp lemon juice
¼ tsp chilli flakes

Break your broccoli into florets. Slice the large florets in half so they cook evenly. Toss them in 2 tbsp of extra virgin olive oil, salt and a generous dose of cracked pepper. Chargrill your broccoli florets on a griddle pan till they are black in some places and lightly cooked through. If you're feeling lazy, you can put all your broccoli on a large grill tray and grill it on high. Just make sure it doesn't char too much.

Lightly toast your almonds in a small, heavy based saucepan.

Prepare your dressing by combining the olive oil, lemon juice and chilli flakes in a small jar. Shake to combine well.

Put your salad together starting with the broccoli in a shallow bowl. Toss through the mint, slithered almonds and dressing. Eat immediately while the broccoli is warm.

Paprika cauliflower salad

Serves 4–6 as a main salad

1 medium-sized cauliflower (around 800g)
Extra virgin olive oil
1 tbsp sweet and smoked paprika
Baby spinach (a few handfuls)
¾ cup toasted pecans

Dressings
1 ¼ cup plain yoghurt
2 tsp sweet and smoked paprika
Extra virgin olive oil
2 tbsp lemon juice
Salt and pepper for seasoning

Divide your cauliflower into florets and chop some in half so they can cook evenly. Place your cauliflower onto a grill tray. Drizzle with olive oil. Sprinkle 1 tbsp. of paprika evenly across the cauliflower and toss through. Cook on high for 15–20 min. Turn each floret once. The top of the cauliflower should be slightly charred and cooked through.

Prepare your dressings. Combine the lemon juice with an equal amount of extra virgin olive oil. Add salt and a generous amount of cracked pepper to taste. Next, combine 2 tsp. of paprika with the plain yoghurt to create a paprika yoghurt. Season generously with salt. Place the baby spinach in the bottom of a salad bowl. Add the chargrilled cauliflower. Drizzle the lemon and olive oil dressing on top. Dollop the paprika yoghurt dressing and serve warm.

Wilted bok choy

Serves 4 as a side dish

10 outer stalks from home grown bok choy, or two whole plants
1 tsp black sesame seeds (optional)

Dressing
1 tbsp white wine vinegar
1 tbsp soy sauce
1 tsp sesame oil

Place the bok choy stalks in a shallow, heatproof dish. Cover with boiling water and let the stalks blanch for 2 minutes. Drain the hot water.

Prepare the dressing by combining the rice wine vinegar, soy sauce and sesame oil in a small jar. Shake well to combine.

Serve wilted bok choy immediately and drizzle the dressing on top. Garnish with black sesame seeds.

Roast cauliflower and quinoa salad

Serves 4 as a main dish

1 whole cauliflower
1 generous glug of extra virgin olive oil
Rind of 1 lemon, finely grated
Salt and pepper for seasoning
1 cup red quinoa (cooked)
3 cups combination of rocket and baby spinach leaves
Handful of chopped parsley leaves

Dressing
¼ cup extra virgin olive oil
1 tbsp red wine vinegar

Chop the whole cauliflower head into florets. Pop them on a grill tray with a generous splash of olive oil and the lemon rind. Season to taste with salt and pepper. Grill for 15–20 minutes or until the tops are starting to blacken then turn off the grill and leave them to continue cooking in their own warmth.

Lay out your salad by placing the baby spinach on a large serving platter. Top with grilled cauliflower and red quinoa. Garnish with chopped parsley.

Prepare the dressing by combining the olive oil, red wine vinegar, salt and pepper in a small jar. Shake well to combine.

Caramelised balsamic broccoli salad

Serves 4 as a side dish

1 head of broccoli (or two large bunches of sprouting broccolis)
½ cup sunflower seeds
4 cups baby spinach
Extra virgin olive oil
Caramelised balsamic vinegar
Salt and pepper for seasoning

Cut the broccoli into florets and lightly steam on the stovetop. Working quickly, prepare your other salad ingredients. Lightly toast your sunflower seeds in a small, heavy based pan.

Scatter baby spinach across the bottom of your salad platter. Place the steamed broccoli on top and then sprinkle the sunflower seeds. Drizzle your platter generously with olive oil and caramelised balsamic vinegar. Grind a little salt over the top then toss gently.

Classic vegetable soup

Serves 4–5

1 onion
2 carrots
Celery tops and heart
Extra virgin olive oil
1.5L vegetable stock
2 cups mixed split peas and lentils (dried)
400g chopped tomatoes
Salt and pepper to season
1½ cups thickly cut shredded red cabbage

Finely dice the onion, carrots, celery tops and heart. Place your finely diced vegetables into a large pot. Add a glug of olive oil and gently sauté.

When the onion is translucent, add the vegetable stock, pea and lentil mix and tomatoes. Simmer for 45 minutes, enjoying the smell wafting through your home as you cook. Test the split peas and lentils to make sure they're all cooked and taste the soup. Adjust your seasoning with salt and pepper. When you're confident the peas and lentils are cooked, add the red cabbage and cook for a further 2–3 minutes.

Middle eastern lemon cake

3 large lemons
¾ cup sugar or honey
6 eggs
2 tsp baking powder
2 cups almond meal
Icing sugar and edible flowers (optional)

Place the lemons in a saucepan full of water. Bring to the boil and simmer for 2 hours, topping up with water if needed. Remove the lemons and let them cool. Do this step the night before if you are short on time.

Preheat the oven to 170°C. Grease a 20cm round springform cake tinand line with baking paper.

Chop the cooled lemons roughly and remove any seeds. Using a stick blender or food processor, pulverise the chopped lemons in a large mixing bowl. Add the sugar or honey, eggs and baking powder. Use the stick blender or food processor again to pulverise the mixture.

Using a spoon, fold in the almond meal until all dried bits are combined. Pour the mixture into the prepared cake tin and bake for 1 hour and 15 minutes, until a skewer comes out clean and the top is golden.

Stand for 5 minutes then remove carefully from the tin. Serve your cake with a dusting of icing sugar and a sprinkling of edible flowers, like pansies.

Lemonade

1½ cups white sugar
½ cup water
2 cups lemon juice (about 16 lemons, depending on their size and juiciness)
4L carbonated water

Create a sugar syrup by combining the sugar and water in a heavy based saucepan on medium heat. Stir as the sugar dissolves in the water. If you can see small air bubbles forming through clear liquid, your sugar syrup is complete.

Add lemon juice to the sugar syrup. Congratulations! You now have lemon cordial that can be stored in the fridge for up to a week.

Make up your lemonade by adding about 1.5cm of lemon cordial to the bottom of a glass and filling the rest of your glass with carbonated water.

Store the lemon cordial and fizzy water separately in the fridge. Create your lemonade at the time when you wish to drink it. If it's a cold day, add warm water to your lemon cordial to create a sweet, tangy warm drink.

Broccoli leaves

Move over kale. Broccoli leaves are the latest in super foods. Use them raw in salads or smoothies. They're great cooked in a stir fry or bundled into cheese parcels. If you're a home gardener, the main difference between broccoli and kale is that broccoli plants deliver broccoli heads as well as tasty leaves.

If you can't get broccoli leaves, don't despair. Most recipes for broccoli leaves are also great with broccoli's cousin kale. Both have super food status, even if kale plants don't work as hard to earn their place in our veggie gardens. Both will help you on your way to glowing health in the depth of a cold winter.

Here are simple recipe ideas for broccoli leaves (or kale) that will inspire you to enjoy this super food for breakfast, lunch and dinner!

Breakfast: Green smoothie

Serves 1

In your blender or smoothie maker combine: 1 banana, ¼ cup blueberries, ¼ cup lemon juice, 1 cup apple juice and 1 large broccoli leaf. Enjoy the goodness!

Lunch: Parsnip, cauliflower and broccoli leaf salad

Firstly, roast the parsnip and cauliflower. Chop the parsnips into 2–3 cm cubes and the cauliflower into florets. Drizzle with olive oil, stir through a clove of grated garlic and sprinkle on some salt. Roast in a medium oven, about 180°C for 30–40 minutes.

Shred six broccoli leaves and place them on a large serving platter. Top with the roasted parsnip and cauliflower. Sprinkle over ¼ cup of lightly toasted sunflower seeds.

Dress the salad with balsamic vinegar, olive oil, salt and pepper.

Snack: Broccoli leaf and parsley pesto

Using a food processor or stick blender, combine 3½ cups shredded broccoli leaves, ½ cup roughly chopped Italian parsley, 1 garlic clove, ⅓ cup fresh walnuts, ¼ cup olive oil and ¼ teaspoon of salt. The mixture should be a thick, creamy texture. Add more olive oil if necessary.

Serve the pesto with carrot and celery sticks as an afternoon snack.

Dinner: Vegetarian ragu with pasta

This is a variation on a traditional pasta sauce made in Tuscany with leafy greens and hand-made, fresh pasta.

In a heavy based saucepan, heat 2 tablespoons of extra virgin olive oil and lightly brown a clove of finely chopped garlic. As the garlic browns, add 1 tablespoon of fresh oregano. The combination of oregano and garlic cooking will fill your home with a hearty smell.

As soon as the garlic starts to lightly brown, add 400g of skinned, chopped tomatoes (a can is fine) and simmer for a few minutes. While the tomatoes are simmering, finely shred 8 large broccoli leaves. Add them to the simmering tomatoes along with ½ cup of cooked lentils. Season to taste with a pinch of sugar, salt and pepper. Let this simmer away for 10–15 minutes while you cook the pasta.

Cook about 400g of pasta, fresh or dry. Combine the pasta and sauce, then serve with Parmesan cheese and olive oil.

SPRING KITCHEN GARDEN

Green shoots transform the garden in spring. Asparagus spears push up through the soil, daring us to snap off the sweet and tender shoots. After a winter of looking hopefully at our snow peas, they're finally ready for crunching and munching. White apple blossom flies into the air with each gust of wind, creating a majestic carpet underneath the trees. It's all beautiful in the garden!

Wholesome greenery is ready to eat in spring and it doesn't mind a mid-spring frost. Lettuce, kale, snow peas, sugar snap peas, broad beans and herbs are all on the menu. If you're lucky, you might also have carrots and beets. Delicious!

These recipes celebrate green in a spring kitchen garden.

Baby broad bean and asparagus pasta

Serves 2 people as a main meal

120g spaghetti
2 garlic cloves
2 tbsp extra virgin olive oil
½ cup baby broad beans
12 asparagus shoots, cut into 3.5cm lengths
2 tsp preserved lemon, finely chopped
2 handfuls of rocket
Shaved Parmesan cheese
Salt and cracked pepper for seasoning

Start by putting a large pot of water with a generous pinch of salt on to boil for your pasta. When the pot is boiling, add your pasta and gently stir the spaghetti to prevent it from sticking to itself.

Meanwhile, prepare the pasta sauce. Lightly sauté the garlic in olive oil then add the baby broad beans, asparagus shoots and preserved lemon. Sauté for 3–5 minutes, season with pepper (the preserved lemon is salty) then take the sauce off the heat. The spaghetti should be ready about now. Stir it through the sauce until it's well combined.

Serve the pasta and sauce in bowls, topping each with a handful of fresh rocket and shaved Parmesan cheese.

Baby carrots with sage

Serves 3 as a side salad

9 baby carrots, tops removed
1 tsp honey
1 glug of extra virgin olive oil
Sage leaves (handful)
Salt and pepper to season

Peel or scrub the baby carrots. Toss them in honey, olive oil, sage leaves and season with salt and pepper.

Pop the carrots under the grill and cook them for about 15 minutes on medium, or until the tops of the carrots are slightly charred. Let them rest a further 10 minutes under the grill to finish cooking through.

Give this recipe a spectacular twist by combining orange, yellow and purple baby carrots.

Spinach pancakes

4 eggs
1 cup wholemeal flour
1 cup milk
1 cup finely chopped spinach
Olive oil for cooking

Place the eggs, flour, milk and spinach into a bowl and mix with a stick blender or food processer until a smooth batter has formed.

Heat one or two heavy based saucepans on a medium-high heat. Add a glug of olive oil and swirl it around so it thinly covers the surface of each saucepan. Add a ladleful of pancake batter to each saucepan. Flip your pancakes when bubbles come up throughout the middle of the pancake.

Serve the pancakes topped with runny poached eggs, micro herbs, a little crumbled feta cheese and dukka.

Spinach pancakes are a great way to use plants that supplied baby spinach leaves in your home garden throughout winter and are now getting old. The leaves are a little tougher and will be best cooked rather than eaten raw.

Radish, asparagus and avocado rice paper rolls

Serves 4 as an entrée

8 asparagus stalks
4 red globe radishes
1 carrot
1 avocado
¼ iceberg lettuce
1 bunch of coriander
1 bunch of mint
½ cup roasted peanuts, crushed
8 rice papers
Chilli flakes

Dipping sauces
Sweet chilli sauce
Soy sauce
Plum sauce (See recipe p. 206)

Prepare the rice paper filling ingredients and place them elegantly on a salad platter. Slice your asparagus stalks across the middle and along the length. Blanch in boiling water for 3 minutes. Thinly slice the radishes with a mandolin if possible. Cut the carrot into matchsticks, matching the length of the asparagus stalks. Slice the avocado thinly, to get the longest possible slices. Finely shred the lettuce. Pick the leaves off from your bunches of coriander and mint. Place your peanuts and dipping sauces in small, individual side bowls.

You will now have an elegant salad platter, with each item deconstructed.

Prepare your rice paper rolls by dipping them in hot water (mind your fingers), lay the 'cooked' rice paper out on a fresh plate. Add your filling to the middle of the rice paper, not too much as you'll need to fold it over. Sprinkle over some chilli flakes and roll it up.

Rice paper rolls are fresh, light and, if you choose, spicy.

Asian broccoli and snow pea salad

2 garlic cloves, finely chopped
1 tbsp finely chopped ginger
1 drizzle of extra virgin olive oil
250g sprouting broccoli
250g snow peas
3 tbsp rice wine vinegar
1 tsp caster sugar
Salt for seasoning

Lightly sauté the garlic and ginger in olive oil in a small pan until the garlic is just cooked. Take care not to overcook. Leave aside to cool a little.

Trim the broccoli to a similar length to the longer snow peas. Place the broccoli in a shallow dish and blanch by covering the broccoli with boiling water for 2–3 minutes. Drain the broccoli stalks. Prepare your snow peas by trimming the hard ends.

Finish the dressing by adding rice wine vinegar and caster sugar to the garlic and ginger in the small pan. Stir until the caster sugar is completely dissolved.

Assemble your salad by tossing the broccoli, snow peas and dressing together in a salad bowl. Season with a little salt. Serve while warm.

INFUSED OILS AND FLAVOURED VINEGARS

Creating a meal with the perfect flavour is like producing a work of art. There's balance between the subtle and the strong… and feeling. Home-made flavoured oils and vinegars are a great way to curate a unique dining experience. They also make lovely gifts.

Rosemary infused olive oil

Take a large sprig of rosemary and place it into a small sterilised bottle. Gently heat ½ cup of extra virgin olive oil until bubbles just start to form on the bottom of the saucepan. Remove from the heat and pour it into the bottle over the rosemary sprig. When the oil has cooled to room temperature it's ready to use. Leave the sprig of rosemary in the bottle so that the flavour builds over time.

Toss potatoes in rosemary infused olive oil and roast in the oven. This will create flavour without the texture from the leaves.

Lemon infused olive oil

Give any dish a citrus tone with lemon infused olive oil. It's perfect in summer salads and for creating a different flavour on braised winter vegetables.

Peel the rind from half a lemon and place it into a heavy based saucepan. Add ½ cup of extra virgin olive oil. Gently heat the lemon rind and oil until you see the first bubbles form on the bottom of the saucepan. Pour the lemon rind and olive oil into a sterilised bottle. Don't panic if your oil looks a little cloudy, this is normal.

Chilli infused olive oil

Turn up the heat in an Asian stir fry or taco bean mix with this chilli infused oil. In Italy, chilli oil is traditionally drizzled on top of some pizzas just before they're eaten.

Finely slice two small chillies and place them into a heavy based saucepan along with ½ cup of extra virgin olive oil. Gently heat the chilli and olive oil until you see the first bubbles form on the bottom of the saucepan. Place a small, whole chilli in the bottom of a sterilised bottle. Pour the heated chilli and oil on top of the whole chilli. Use with caution as the flavour intensifies with time and as the ratio of chilli to oil decreases when you use the oil.

Raspberry vinegar

Fruit vinegars are fantastic in most salad dressings. They're sweet, sour and full of fruity goodness.

Place ¼ cup of fresh or frozen raspberries and 1 tablespoon of caster sugar into a sterilised glass jar. Pour 1 cup of white wine vinegar over the berries and sugar, mixing gently to combine. Seal the lid and store in a dark cupboard at room temperature for at least one week, giving the mixture a shake every few days. Remove the mixture from the cupboard and strain out the raspberries. Now you have raspberry vinegar.

If you've mastered raspberry vinegar, try strawberry or blueberry vinegar. The method is the same, just different fruit and a new flavour.

Caramelised balsamic vinegar

Nothing finishes an amazing salad like caramelised balsamic vinegar. Sweet and savoury all in one! Here's how to make your own at home.

Bring 3 cups of balsamic vinegar to the boil. Stir through 2 cups of brown sugar then reduce to a simmer. Simmer for 30–40 mins, until the mixture has reduced by half. Place in a sterilised bottle.

Drizzle on figs or strawberries as a dessert. Finish a French lentil salad with caramelised balsamic vinegar on top.

SUMMER KITCHEN GARDEN

Summer is the season of abundance in the kitchen garden. It's when the fruits of gardening are best enjoyed. There's plenty to pick, eat and share with friends. The flavours are strong. The weather invites us to be outside, especially as the sun goes down and warm days give way to balmy nights.

Kitchen garden cooking doesn't need to be complicated. With the freshest produce, it's all about bringing out flavour created by the sun and soil in your garden.

Heirloom tomato salad

Serves 4 as a side salad

Tomatoes taste best straight from the garden, without refrigeration. This recipe lets heirloom tomatoes from your kitchen garden shine with its simplicity.

2 large beefsteak tomatoes
2 large black Russian tomatoes
2–3 pinches of salt
2–3 pinches of caster sugar
1 drizzle extra virgin olive oil
Black pepper

Thinly slice your tomatoes to show off the gorgeous inside patterns. Sprinkle the slices with a few pinches of salt and sugar. Display the slices on a white plate and drizzle with olive oil and freshly ground black pepper on top.

Serve with crusty bread and fancy cheese.

Raspberry fool

Serves 4

250ml pouring cream
250ml plain yoghurt
½ cup icing sugar
1 lemon (juiced)
500g raspberries

Whip the cream until it holds its shape in the bowl. Stir through the yoghurt, icing sugar and lemon juice until well combined. Fold the raspberries through the cream and yoghurt mixture, reserving 8–10 for decorating.

Serve in a wine glass topped with the reserved whole, fresh raspberries and a sprinkle of caster sugar. You could also take the raspberry fool on a picnic by putting it into small glass jars.

Roast garlic and pumpkin tarts

Makes 12 small tarts

4 small garlic bulbs
250g pumpkin
1 glug of extra virgin olive oil
6 free range eggs
250g ricotta cheese
100g feta cheese
3 sheets of store-bought shortcrust pastry
Salt and pepper for seasoning

Prepare the garlic by finely chopping off the bottom of the bulb, so that each clove is sliced across the bottom. Prepare the pumpkin by chopping it into 2cm cubes. Place the garlic bulbs and pumpkin cubes on a roasting tray, drizzle with olive oil and season with salt and pepper. Toss with your hands to combine the oil, seasoning and vegetables. Roast at 180°C for about 40 minutes.

Combine the eggs and ricotta in a large mixing bowl.

Prepare the tart moulds. Take a 12-tart tray (or muffin tin) and thoroughly grease each mould with olive oil. Take the shortcrust pastry out of the freezer and let it defrost for about 5 minutes before you work with it. Use a butter knife to cut the pastry into circles — four per sheet. Place each into the tart moulds.

Spoon the egg and ricotta mixture into the pastry bases.

Gently squeeze at least two small cloves of garlic into each tart then place at least two pumpkin pieces into each tart. Finally, crumble a little feta cheese onto each tart and push it down, so that it sinks in. Cook at 200°C for 20 minutes or until the pastry is cooked and the tart is set through.

Serve the tarts with a light, leafy green salad.

To make the most of your shortcrust pastry, roll the pastry 'scraps' into lengths and cook on a greased tray at the same time as the tarts.

Yoghurt and cucumber soup

Serves 4 as a starter

1.2kg cucumber (4 large Lebanese cucumbers that 'got away' in the garden are perfect)
1½ cups plain or Greek-style yoghurt
1 garlic clove
3 tbsp of lemon juice
⅓ cup dill (roughly chopped)
⅓ cup combination of tarragon and chives (roughly chopped)
Salt and pepper to taste
¼ cup extra virgin olive oil
mint

Combine all ingredients in a food processor, blend well. Chill for at least 2 hours.

Check seasoning immediately prior to serving. Garnish with mint.

Zucchini

Summer is the season where zucchini plants know no boundaries. Zucchini flowers and their fruit just keep coming and coming… and coming. Ooops, I missed a zucchini and now it's a massive marrow. What am I going to do with the generosity of this blessing?

Having a garden glut is a good thing. Better than not having enough. The best thing for a glut of zucchinis is inspiration.

Zucchini Chutney

1.8kg zucchini, cut into 1–2cm cubes
6 onions, finely diced
500g white sugar
1 tbsp table salt
½ tsp curry powder
½ tsp ground turmeric
2 tsp mustard powder
2 cups white vinegar

Combine all ingredients in a large, heavy based saucepan. Bring to a gentle boil for about 1½ hours, stirring every now and then to prevent the chutney sticking to the bottom of the saucepan. Keep the lid off so moisture can escape. Your chutney is ready when the zucchini has turned golden and the consistency is to your liking. If your chutney is too watery, keep it boiling a little longer.

Bottle in sterilised jars.

Zucchini and bean salad

Serves 2 as a main or 4 as a side salad

3 large handfuls stringless beans (yellow, green and purple)
3 small zucchinis (yellow and green), about the length of your hand

Dressing
3 garlic cloves
½ bunch mint leaves
½ cup extra virgin olive oil
½ lemon (juiced)
Salt and pepper for seasoning

Top and tail the beans. Chop the zucchini into lengths around the same size as the beans. Pop the zucchini and beans into boiling water along with a pinch of salt and boil for 5–6 minutes to cook. Drain all the water off. The zucchini should be soft, but still retain its shape.

Prepare your dressing by finely chopping the garlic and picking off each mint leaf from the bunch. Combine the dressing ingredients in a small jar and shake well to combine.

Combine the hot, just cooked zucchinis and beans with the dressing. Leave the salad at room temperature for at least 30 minutes so that the vegetables have time to absorb the flavour.

Perfect with crusty bread as a simple lunch.

Raw zucchini and lemon salad

Serves 2 as a main meal or 4 as a side salad

2–3 young zucchinis
1 lemon (juiced)
50g Parmesan cheese, thinly sliced and crumbled
1 drizzle of extra virgin olive oil
Salt and pepper

Using a vegetable peeler, slice the zucchinis into ribbons. Combine the zucchini ribbons, lemon juice, Parmesan cheese and olive oil in a salad bowl. Season to taste with salt and pepper.

Zucchini stuffed with goat's cheese

Serves 4 – 6 as a side

2 medium zucchinis
100g goat's cheese
Handful of mint leaves, chopped
1 drizzle of extra virgin olive oil
Salt and pepper to season

Make a slit lengthwise down the middle of your zucchinis and stuff them with goat's cheese. Pop them onto individual pieces of lightly oiled aluminium foil. Sprinkle the zucchini with fresh, chopped mint, salt and pepper. Drizzle with olive oil.

Wrap up the foil so the zucchinis are fully encased and bake in a hot oven, about 220°C, for 30 minutes. Your zucchinis will be tender when they're ready. Try this one on the BBQ or a campfire, baking the zucchini on hot embers.

Vegetarian kebabs

8 bamboo kebab skewers
3 medium zucchinis
1 red onion
400g haloumi cheese
3 red capsicums

Marinade
1 large glug of extra virgin olive oil
½ lemon (juiced)
Salt and pepper

Soak the bamboo kebab skewers in water for at least 20 minutes to prevent splinters from emerging.

Prepare the kebab ingredients. Chop the zucchinis into 2–3 cm round chunks. Chop the onion into four quarters and separate the quarters so that you have at least two rings in each chunk. Dice the haloumi into 2 x 2cm cubes and slice the capsicum into chunks of similar size to the onion.

Prepare the marinade by combining all ingredients in a mixing bowl. Toss the kebab ingredients through the marinade and let it sit for at least 30 minutes.

Thread your kebab ingredients along each kebab and cook on the BBQ, turning regularly.

Enjoy these gorgeous vegetarian kebabs that are light, tasty and good for you!

Zucchini 'pasta'

Serves 2 as a main meal, 4 as a side

2 garlic cloves
1 glug of extra virgin olive oil
1 handful of oregano leaves
400g chopped tomatoes
Salt, sugar and pepper for seasoning
2 large zucchinis
Shaved Parmesan cheese to serve

To make your pasta sauce, finely chop the garlic and sauté in olive oil until just cooked. Add the oregano leaves and tomatoes. Let the sauce simmer for at least 10 minutes. Season with salt, sugar and pepper to your liking.

Meanwhile, bring a pot of salted water to the boil. Using a spiral slicer, create thin zucchini ribbons that look like spaghetti. Plunge the zucchini ribbons into the boiling water for 30 seconds to blanch and then drain the ribbons.

Serve the zucchini spirals in little mounds, topped with the tomato sauce and a generous helping of shaved Parmesan cheese.

Grilled zucchini and capsicum

Serves 4 – 6 as part of a starter platter

3 large red capsicums
3 large zucchinis
1 large drizzle of extra virgin olive oil
Salt and pepper for seasoning

Slice the capsicums in half lengthwise and remove their seeds. Put them under a hot grill until they char, about 20 minutes. The more charred, the easier they will be to peel and the more flavoursome. Remove them from the grill and stand them aside in a bowl to sweat and cool. After they're cool enough to touch, peel their skins. Slice the capsicums lengthwise into thin strips.

Slice the zucchinis lengthwise into pieces that are about 3mm thick and a similar length to your capsicums. Brush with olive oil and grill for about 10 minutes or until they're charring slightly. Allow them to cool then combine the grilled zucchinis and grilled capsicum.

Serve immediately as part of an antipasto, with a slight drizzle of olive oil and some salt and pepper. Alternatively, store them in a jar in the fridge for up to a week with olive oil, salt, pepper and some chopped garlic.

Zucchini muffins

Makes 12 muffins

3 free range eggs
1 large zucchini, grated (approximately 600g/ 3 cups of grated zucchini)
1 cup olive oil
1½ cups wholemeal flour
3 tsp baking powder
1 tsp ground cinnamon
⅔ cup brown sugar
½ cup sunflower seeds
1 cup blueberries (fresh or frozen)
A few handfuls of rolled oats (optional)

Lightly beat the eggs in a large mixing bowl. Add all the other ingredients and combine well.

Line 12 muffin moulds with patty pans and fill with the mixture. If desired, sprinkle a few oats over the top of each muffin. Bake in a the oven at 180°C for 20 minutes.

This is a great lunch box regular. Tasty and brilliant for sneaking in some extra vegetables.

Pesto

Intense flavour and fresh, green goodness. There's so much to love about home-made pesto. It's easy to make, flavour combinations can be wild, and a dollop of home-made pesto transforms mundane to gourmet.

The possibilities for home-made pesto are limitless. Experiment with leafy greens you've got in the garden and locally grown nuts to create a unique flavour sensation that's bursting with goodness.

Italian pesto

The original flavour sensation is made with basil, toasted pine nuts, Parmesan cheese, garlic, lemon juice, salt and olive oil. Pesto was made in Italy long before food processors, with ingredients combined using a pestle and mortar. It's a perfect way to bottle up summer's goodness. Italian pesto is great with pasta and on top of minestrone soup.

4 cups basil leaves
¼ cup toasted pine nuts
⅓ cup extra virgin olive oil
50g Parmesan cheese
1 garlic clove
1 tsp salt

Combine all ingredients with a food processor or mortar and pestle. Bottle in sterilised jars and store in the fridge.

Parsley and walnut pesto

3 cups flat leaf parsley leaves
⅓ cup extra virgin olive oil
⅓ cup walnuts
1 garlic clove
1 tsp table salt

Combine all ingredients with a food processor or mortar and pestle. Bottle in sterilised jars and store in the fridge.

Parsley and walnut pesto is perfect with spaghetti. It has a stronger, earthier flavour than traditional pesto. It's also great thrown in dollops on top of a chickpea salad.

Coriander and lemon pesto

2 cups coriander leaves and chopped stalks
½ lemon, zested
1 kaffir lime leaf
1 garlic clove
¼ cup macadamia nuts
2 tbsp extra virgin olive oil
1 tsp table salt

Combine all ingredients with a food processor or mortar and pestle. Bottle in sterilised jars and store in the fridge.

Coriander and lemon pesto is perfect with fresh rice paper rolls. It's also great with leafy green salads.

Carrot leaves, basil and macadamia pesto

If you've got carrots in your kitchen garden or at the local famer's market, then you'll have access to baby carrot leaves. Baby carrot leaves are soft and tender. At home, you can harvest a few from each plant, leaving the carrots in the ground to continue growing. Don't use the leaves if they've turned stiff as they'll make your pesto taste bitter and give a harsh texture.

Carrot leaves lighten the flavour of a traditional basil pesto. This makes it better suited to stirring through a lentil salad or for adding that special burst of flavour to your lunch time sandwich.

1½ cups baby carrot leaves
1½ cups basil leaves
¼ cup extra virgin olive oil
⅓ cup macadamia nuts
1 garlic clove
1 tsp table salt

Combine all ingredients with a food processor or mortar and pestle. If the texture is too chunky, add a little extra olive oil. Bottle in sterilised jars and store in the fridge.

Herb butter

100g salted butter
2 tbsp finely diced seasonal herbs
1 garlic clove, finely diced

Select your seasonal herbs from the garden. Any combination of parsley, tarragon, basil, oregano and coriander will work well. Experiment with what you've got available.

Let your butter soften by placing it at room temperature for 30 minutes.

Combine the butter, herbs and garlic in a mixing bowl by mixing extremely well. Store your herbed butter in a small jar in the fridge.

Use the butter to toss through steamed corn or green beans.

Plum sauce

1.5kg plums, washed and roughly chopped
1½ cups white vinegar
1 cup white sugar
1½ tsp table salt
1½ tsp ground Chinese five spice
1 small red onion, finely diced
1 tbsp ginger, finely diced

Place the chopped plums in a heavy based saucepan. Add all the other sauce ingredients and bring to a gentle boil. Boil for about 30 minutes or until the plums have fallen apart.

Take your plum mixture off the boil and whiz it with a stick blender or food processor.

Bottle your sauce in sterilised jars.

Gift to neighbours and store in the fridge.

Select plums that are overripe for maximum sweetness and aroma. Plums make an amazing sauce, just perfect for many Asian dishes. Homemade plum sauce is the perfect accompaniment to Vietnamese rice paper rolls or a vegetable stir fry.

Tomatoes and passata

Making passata is a family tradition because it's best done in big quantities. There can be mess all over the kitchen — tomato seeds, basil stalks and garlic skin — and a delicious smell wafting through the house, along with the sounds of your favourite Italian opera. Passata captures the flavours of summer in a bottle, each one bringing a richness and sweetness to cooking in the colder months.

The quantities in this recipe are designed for home gardeners with a modest crop. It can be easily tripled or quadrupled if you decide to buy tomatoes in bulk from a commercial grower.

12 medium sized tomatoes
6 garlic cloves
5 large sprigs of basil
2 tbsp extra virgin olive oil
1 small red chilli
1 teaspoon of cooking salt
½ teaspoon of caster sugar

Step 1. A soundtrack of Italian opera is the secret, unwritten ingredient for incredible passata. Before you get started, find your favourite album and turn up the volume.

Step 2. Prepare the tomatoes. Score the top of each tomato with a cross and plunge them into boiling water for about 10 seconds. Pull the tomatoes out one by one. As they come out, peel off the skin. Place peeled tomatoes into a bowl.

Step 3. Get the flavour happening. Peel and finely chop the garlic cloves. Remove basil leaves and finely chop the stalks. Place the finely chopped garlic and basil stalks into a heavy based saucepan along with the olive oil. Cook on a medium heat until the garlic has just browned.

Step 4. Create your passata. Add the peeled tomatoes to the saucepan then turn the heat up a little. Let the whole tomatoes cook for at least 10 minutes. Stir the pot to make sure the garlic and basil stalks don't brown. Take the tomato mixture off the heat and use a potato masher to crush the tomatoes to create a sauce-like consistency. There will still be small chunks of tomato, creating passata with character.

Step 5. Finish off your flavour. Return the passata mix to a medium heat and add in the whole chilli, salt and sugar. Lightly boil for up to 30 minutes to reduce the liquid and intensify the flavour. Finely chop the basil leaves and add to the passata. After a few minutes, test the seasoning and add more salt or sugar as needed.

Step 6. Bottle up the goodness. Place hot passata into sterilised bottles. Re-purposed bottles are fine, just sterilise them using your oven or microwave first.

Share your passata bottles with the people who like to listen to your favourite Italian opera.

STONE FRUIT HARVEST

The sweet, aromatic flesh of apricots, plums, nectarines and peaches are a feast for the senses. Summer's the time for this gorgeous fruit to weigh down the branches of backyard trees. Time for harvest.

A plentiful stone fruit harvest will produce much, much more than one household can eat. It will also produce some fruit with small blemishes. Here are five ideas for making the most of your bounty.

1. Eat

The very best of your harvest is for eating fresh. Pick stone fruit a few days before it's ripe. Stone fruit develop their flavours beautifully in a fruit bowl, reducing the likelihood that birds get more than their fair share.

2. Poach

Fruit that's fallen on the ground or is blemished are perfect for poaching. Simply cut out the imperfections when preparing your fruit. If you have more than one type of stone fruit that's ready for harvest, put them all in together.

Step 1. Prepare your fruit by washing them, removing blemishes and evenly slicing them. There's no need to peel stone fruit. Weigh your sliced fruit so you know how much sugar to add in the next step.

Step 2. Place your sliced fruit in a heavy based saucepan. Add 2–3 tablespoons of brown sugar for every 500g of sliced fruit. Add spices to your personal taste. You might add a single cinnamon quill, fresh nutmeg, cloves, ground cardamon or a combination of spices. Do not add water.

Step 3. Gently bring your fruit, sugar and spices mix to a simmer. Stir your fruit as the mixture heats to prevent it sticking to the saucepan and burning. The fruit will have enough moisture in it to create its own delicious poaching liquid. Poach for 10–15 minutes until the fruit is cooked and the flavours have infused.

Step 4. Serve your poached stone fruit with yoghurt and toasted nuts. Freezing poached fruit is also a great way to keep the taste of summer alive when the weather gets cooler.

3. Jam

All stone fruit can be made into jam. This recipe is for apricots from your home tree, including those that are slightly blemished or have fallen on the ground.

Step 1. Prepare the fruit by washing, removing blemishes and chopping them into 2 x 2cm chunks. Removing all of the dirt is important as little bits of dirt will make your jam foam. There's no need to peel stone fruit. Weigh your chopped apricots. A good amount for a batch of jam is between 500g and 1.5kg. Add your chopped apricots to a large, heavy based saucepan.

Step 2. Weigh your white sugar. You'll need equal weight of sugar and chopped apricots. Place the white sugar in the saucepan with the apricots.

Step 3. Turn up the heat. Gently stir the sugar through the apricot pieces so that the apricots are macerated. This prevents the sugar and apricots burning. Bring the mixture to the boil. Hold on rolling boil for 20–30 min, enough to break the shapes in the apricot pieces so they're nice and mushy to spread on your toast.

Step 4. Take the apricots off the heat and skim off any white foam. Break up any large pieces with a potato masher. For a thick jam, add pectin and bring your jam gently back to the boil for 5 minutes. If you don't stir as you bring your jam back to the boil, it will stick slightly to the bottom of your saucepan and caramelise, creating a lovely deep golden colour. Check that your jam has set by removing a little on a spoon and letting it cool. It should cool to a gluggy consistency.

Step 5. Bottle your jam in sterilised jars.

4. Dry

The hot summer sun has been used to dry fruit for more than a thousand years. The timing is perfect. Most fruit ripens in summer when the sun is at its hottest. To dry fruit in the sun requires a little patience and supervision. You can also dry fruit on a tray with baking paper in an oven (on low) or in a purpose-built dehydrator.

Step 1. Wash, chop your fruit in half and remove stones. Slice your fruit so that each piece has an even thickness.

Step 2. Place your fruit on a drying rack in a purpose-built dryer and run the dryer on high overnight. Test your fruit in the morning. If it's not yet dry enough for your liking, keep the dryer running. If you want to store your fruit for more than a few days, you'll need to reduce the moisture content to be roughly consistent with dried fruit that you purchase from a store. Consume homemade dried fruit within a month — delicious!

HOME-MADE KITCHEN

BREAD

The smell of freshly baked bread gives us a feeling of warmth and comfort. All is good in the world when there's fresh bread ready to be eaten.

Bread has been a staple food for more than a thousand of years of human history. Folklore that celebrates the grain harvest and bread making pops up everywhere, from Scotland to South America. In Norway, it's believed that boys and girls who share the same loaf a destined to fall in love and marry.

Here are three great bread recipes.

Classic Seed and Nut Loaf

This recipe is easy and turns out spectacular bread every time. The secret? It uses yeast and doesn't require any kneading, just a little planning as the dough rises overnight.

1½ cups wholemeal flour
2½ cups plain flour
1 tsp yeast
2 tsp salt
2¾ cups water
1 cup seeds and nuts (any combination of: linseeds, pepitas, poppy seeds, sesame seeds, sunflower seeds and flaked almonds works well)

Mix all ingredients together, to make a wet, sticky dough. Leave for 10–24 hours covered. Don't knead. It will triple in size (approximately) and look sticky and holey.

Plop the wet, sticky dough into a large, greased baking dish with a lid. If you'd like to, sprinkle some seeds on top to give the bread some pizzazz.

Put it into a hot oven (about 240°C) for 30 minutes.

Remove the lid, then bake for a further 20 minutes or until brown and crusty.

Leave the bread to cool on a rack.

Sourdough

There's art and passion in soughdough making. There are also workshops from cooking schools and a long list of books about how to make it. In addition to the usual bread making, sourdough requires you to maintain a starter culture, make time for resting it and for fermenting.

Here's an overview of the steps. They're designed for someone new to get an idea of what's involved, or for someone experienced to be re-inspired to get baking again.

Sourdough starter
60g rye flour
400g white wheat flour (or white spelt flour)
240g rainwater
1½ tsp finely ground sea salt

Step 1. Grab some starter culture from a friend, a shop or online. This will be a sticky, bubbly combination of rye flour, water and culture in a glass jar. If you're feeling wild, make your own by exposing a mixture of rye flour and rainwater to the air for a few days, just be prepared for this to take a few goes to get right.

The starter culture can live in your fridge, just take out a little each time you bake. You'll also need to 'feed' the culture with little bits of extra rye flour and water each time you take some out.

Step 2. Wake up your starter culture. Bring it to room temperature. Combine the 50g of starter with rye flour and 90 ml water. Leave in a warm place for 6–12 hours until it becomes active. You'll know it's active because it will increase in size and look either foamy or spongy, with bubbles throughout.

Step 3. Mix the activated starter with other bread ingredients or flour, rainwater and salt.

Step 4. Rest the dough for 15–20 mins. Air-knead for 5 minutes. For the uninitiated, air-kneading involves throwing your dough into the air and then slap/throwing it onto your bench. It develops elasticity.

Step 5. Let the dough rise for 4–6 hours at room temperature (or overnight if it's cold). You can use a wicker basket lined with a lightly flowered tea towel to give the dough a nice shape. Spray a mist of water over the dough and sprinkle on sesame seeds before it rises. The dough should roughly double in size.

Step 6. Bake at 235°C for 10 minutes then reduce to 215°C for a further 25–30 minutes. If you're unsure whether the loaf has cooked through, tap on the side. Cooked loaves have a firm crust and a hollow sound.

Wholemeal flatbread

Flatbreads are a staple in Middle Eastern and Mediterranean cooking. This recipe goes beautifully with a Turkish feast. Simply top the flat bread with a generous helping of beetroot dip and spicy lamb mince. Garnish with mint leaves.

1 cup self-rising flour
1 cup wholemeal flour
1 cup Greek style yoghurt
1 tsp salt
2 tbsp extra virgin olive oil (plus extra for brushing)

Combine all ingredients and form a ball then turn out onto a lightly floured surface and knead for 2 minutes until smooth. Cover the dough and let it rest for at least 15 minutes.

Divide the dough into 4–6 equal pieces and roll out on a lightly floured surface into rough oval shapes. You'll need them to be about 5mm thick. Brush each flatbread with a little olive oil.

Bring a chargrill pan or heavy based saucepan to a medium heat. Add flatbreads individually and cook for 5 minutes, turning once, until charred and cooked through.

HERBAL TEA

Hold a hot cup of tea in your hands. Take a sip and let it warm your body from the inside. Tea is comfort to the soul.

Making your own herbal tea straight from the garden is surprisingly simple. Home grown tea gives you even more reasons to feel good about taking that sip. There are zero food miles and zero waste from your humble brew. To make each pot of tea, take a small handful of leaves, fresh from your garden. Place them into a pot of boiling water and let them steep for 3–5 minutes. Take your first sip. Enjoy!

Lemon verbena

A rich, lemon flavoured tea. Lemon verbena leaves can be used fresh from the plant or dried. It's known for being a mild sedative and for help with digestion.

Lemon verbena plants like full sun and rich soil. They're frost sensitive, so it's best to grow them in a pot or in a frost-free part of your garden. The plant grows into a small shrub.

Basil

Sweet, aniseed flavoured tea. Basil comes in many shapes and sizes, all suitable for tea. There's sweet basil that's known for its essential role in Italian cooking. Thai basil that helps to create the unique Southeast Asian flavour. Then there's India's holy basil, or Tulsi, scientific name 'Ocimum sanctum'. Holy basil is known for its role in Ayurvedic medicine, where it's used to treat a wide variety of conditions from the common cold to digestive complaints. Holy basil is the best-known basil tea.

All the basils like full sun and rich soil. They're frost sensitive and turn overnight from lovely leaves into a brown shrivelled mess. So, pick your basil before the frost hits and dry it inside for your winter brew.

Mint

Fresh, invigorating and… well, minty. Mint is a classic Moroccan tea. People drink it as a pick-me-up, to sooth an upset stomach and as a cure for bad breath.

There's quite some diversity in the type of mints available. Think spearmint, Vietnamese mint and classic mint. The 'classic winter mint', with its dark green leaves, produces the best flavoured tea. Mint likes a spot with rich soil and full sun/ part shade. It sets out runners and can take over your garden in the summertime, dying back over the colder months. To contain your mint, grow it in a pot or away from your veggie patch.

Lavender

Strong, yet soothing, lavender tea is known for helping you to relax and unwind. The distinctive lavender smell translates into the tea's sweet flavour.

While different lavender species have a similar smell, their chemical compositions are distinct. Try English lavender for a soothing tea. Harvest mature flowers to put into your tea.

Lavender prefers full sun and rich, well-drained soil; but will tolerate part-shade and rocky clay soil. It's not frost sensitive, so grow it anywhere in the garden.

Lemongrass

A zingy lemon tea that takes you to Asia. In the herbal world it's known for a vast range of beneficial properties, including lowering cholesterol, improved digestion, healing the common cold and reduced arthritic pain.

Lemongrass grows best in full sun, with rich soil. It's frost and cold sensitive, so grow it in a pot and bring that pot inside if you live in a cold climate.

Bay leaves

The deep, rich flavour of bay leaves makes them the perfect brew following a big meal. Bay leaf tea is known for soothing digestion and improving gut health.

Bay leaves come from a bay tree that can grow up to 5 meters tall. So, if you plant a bay tree, you'll have tea absolutely any time! You'll also have bay leaves for cooking and for gifting to friends. Most people prune their bay trees and use them in hedging or as a topiary garden feature. Bay trees like full sun and rich soil. They're sensitive to frost but grow just fine in a frost sheltered garden position or in a pot near your home.

Lemon fruit

A few thick slices of lemon in a large cup of warm water is the way start each day. There's a zing and a lift from the lemon, a good start to your day's hydration with the water and warmth that kick starts your day. Lemon tea is known as an excellent detox. There are even celebrity endorsements and a special lemon tea detox diet — lemon tea is famous.

Lemons trees are frost sensitive and like rich, well-drained soil, as well as full sun. Lemons grow well in pots and fruit throughout the wintertime.

Rosemary

Fresh and stimulating, with depth of flavour, rosemary tea is made with the leaves and twigs, either dried or fresh. Rosemary tea is known for improving circulation.

Rosemary prefers full sun and rich soil, but will tolerate part-shade and rocky, clay soil. It's a perennial and is frost tolerant.

Blending your tea

If you're really into tea, experiment with combining your home-made goodness. Try lavender and bay leaves for a strong and soothing combination, mint and lemon verbena for an uplifting tea or lemon rind and rosemary for a rich and zingy mid-morning cuppa.

Dried tea leaves make a beautiful gift, especially if you package them in a glass jar with a ribbon.

PRESERVE A WILD HARVEST

There's wild food that's ripe for the taking in the summertime. Some urban streets are lined with plum trees, many parks have dandelion weeds with bright yellow flowers and blackberries are often found growing near waterways on the urban fringe. Make the most of a wild harvest by eating some and preserving the excess as gifts or for your cupboard as a memory of these warm, summer days.

Dandelion and kale sauté

Dandelions are easy to see in a lawn when they haven't been mown over. They give zing to your regular leafy green sauté.

Step 1. Harvest a bunch or two of fresh, new dandelion leaves. Don't pick the older leaves, they'll have a stronger bitter taste. Not sure if you can identify a dandelion? Harvest an equal weight of kale from your garden.

Step 2. Roughly chop your dandelion leaves and kale then sauté them on a medium heat with a little extra virgin olive oil. When the leaves are nearly sautéed, move them to one side in the pan and in the other side of the pan you can drizzle a little more olive oil. Pop in the garlic and oregano and cook for 1 minute until just cooked. Combine the garlic, oregano and leaves in the saucepan And season with salt and pepper.

Step 3. Serve your sautéed leaves with crumbled feta cheese as a side dish. If you've got too much, let your sauté mixture cool and preserve it in your freezer.

Plums in syrup

There are plums, small and large, on urban street trees in some cities. Depending on the type of plum, they can be ready for harvest as early as late spring – mid-summer. Don't be shy about picking fruit before it's completely ripe. You can let fruit ripen a little at home. You can also preserve fruit that's nearly but not quite ripe. Here's the easiest way to make the most of a wild plum harvest.

Step 1. Wash your plums. If they are large enough, slice them in half and remove the stones. Discard or remove sections that are damaged by grubs. Let your plums dry off.

Step 2. Make your sugar syrup. Depending on the ripeness of your plums, consider different ratios of sugar to water. For a light syrup, combine 1 litre of water with 1 cup of white sugar. For a heavier syrup, combine 1 litre of water with 2 cups of white sugar. Combine the water and sugar in a heavy based saucepan, bringing it gently to the boil. Stir until all the sugar is dissolved.

Step 3. Prepare your preserving jars by sterilising them.

Step 4. Make the preserve. Place your plums into the jars then cover with the sugar syrup. Seal the lid tightly then place the jar into an oven at 100°C for 35 minutes.

You can preserve other stone fruit, like apricots or sour cherries, using the same syrup method.

Blackberry jam

Wild blackberry picking is a true family experience, popular with children who often eat as much as they put into their buckets. Check with your local council to make sure the blackberries were not recently sprayed with herbicide.

Blackberries
Sugar (white or jam setting sugar)
1 lemon (juiced)

Step 1. Pick your blackberries, give them a wash and let them air dry.

Step 2. Weigh the blackberries. You'll need an equal weight of white sugar to the weight of berries. Use jam setting sugar if you're keen for a stiffer jam.

Step 3. Combine the blackberries and sugar in a heavy based saucepan. Add the juice of a lemon for every 1.2kg of berries. Bring to the boil then boil for 15–20 minutes until the mixture thickens. Just before you take the blackberries off the heat, crush them with a potato masher.

Step 4. Prepare your preserving jars by sterilising them. Place the blackberry jam into sterilised jars.

If you don't have access to wild blackberries, this recipe works well with other berries or a mixture of berries.

Yoghurt

Delicious, creamy and healthy. Yoghurts have been enjoyed by people around the world for thousands of years. Mongolian herders make Tarag from the milk of cattle and yaks; in Africa there's the buttermilk-like Amasi; and across the Mediterranean there's a thick yoghurt made from the milk of cows, sheep and goats.

Shops are quick to sell yoghurt making gadgets. Don't be fooled. It's easy to make yoghurt at home. Yoghurt is an ancient tradition that doesn't need specialist equipment.

1L milk
1 tbsp pot set yoghurt

Step 1. Heat the milk in a saucepan so that it's nearly boiling or at 92°C. You'll notice little bubbles slowly coming to the surface. If the milk boils, don't worry, it doesn't affect its yoghurt making properties it just makes a mess in your kitchen. Heating milk changes its protein. You can use full cream, skim or unhomogenised milk. In fact, any milk that's come from an animal will be fine. Full cream milk gives the thickest, creamiest finish.

Step 2. Let your milk cool down, so that it's just above body temperature. We test our milk's temperature using the highly scientific method of putting in a finger and counting to three. The milk should be hot enough to make you want to remove your finger at the count of three. It should be warmer than your body temperature. We also guess the milk's temperature in relation to the time that it's had resting. For 1L, 5 minute's rest after step 1 is about right. For those who like to be precise 35°C–40°C on a kitchen thermometer.

Step 3. Take 1 tablespoon of cultured yoghurt and vigorously stir it through the hot milk. Using a whisk will make sure the cultured yoghurt and milk are fully combined and there are no 'chunks' of yoghurt.

Step 4. Pour the hot milk and cultured yoghurt mixture into 1-2 large sterilised glass jars. Attach the jar's lid.

Step 5. Wrap the sealed jar in an old towel so that it doesn't cool down too quickly. Let it sit at room temperature for 5–12 hours. The yoghurt is ready to refrigerate when the milk has thickened.

Step 6. Refrigerate for up to 6 weeks.

From this basic yoghurt recipe you can make anything. Greek yoghurt is made extra thick by straining excess water from the basic yoghurt recipe. Sweet strawberry yoghurt can be 'pot set' by including strawberry jam in the bottom of the jar before step 4, where your pour the combined hot milk and cultured yoghurt into a jar. Alternatively, stir through jam, honey and cinnamon after your yoghurt has set.

Almond milk

Silky smooth and creamy. Fresh and light. Local. It's homemade almond milk.

Before refrigeration, almond milk had two advantages over cow's milk. Firstly, it lasts longer than unpasteurised cow's milk at room temperature. Secondly, you can make the quantity that you need from a dry store of whole nuts.

To make almond milk you'll need nuts, water, a blender and some cheesecloth to strain the milk. You might also like to sweeten the milk with maple syrup, honey or sugar syrup.

1 cup almonds
3 cups filtered water
2 tsp maple syrup

Step 1. Soak the almonds overnight. Soaking almonds makes it easier to extract the creamy milk.

Step 2. Rinse the soaked almonds and add to the blender with the water. Blend until the almonds are finely ground.

Step 3. Strain the almond mixture through some cheesecloth. Massage the almond mixture until you have all the milk that you can squeeze out.

Step 4. Make it sweet. Add the maple syrup and stir.

Step 5. Chill in the fridge for 30 minutes.

Use your homemade, artisanal almond milk in a dairy-free smoothie. Infuse it in a pot of chai. Savour a long, cold glass without any distractions.

After separating out the milk, you'll be left with almond meal which is perfect for home baking. Why not try baking a Middle Eastern lemon cake to use your almond meal.

LOVE LETTER TO PLANET EARTH

Dear Planet Earth,

Thank you for providing an incredible place for me, and my fellow humans, to live and thrive.

I love your snow-topped mountains, their clear and bubbling streams, and snow gums that glisten after rainfall. I love your deep oceans and marvellous diversity of colourful creatures living in coral reefs, your sandy beaches and white sea eagles that soar above ocean cliff tops. I love your forests of spotted gums, wild cycads and bower birds that collect bright blue objects.

When breathing deeply to inhale fresh, clean air — I thank you. When my thirst is quenched with clear, sweet water— I thank you. When I eat ripe fruit and crisp vegetables in their season — I thank you.

I am sorry, oh so very sorry, for the scarring left behind by me and my fellow humans upon your beautiful surface. I weep for the beautiful places and landscapes that have been destroyed, for the pollution that wreaks havoc with every life form sustained by you, and for the climate that is rapidly changing.

I wish to leave this life knowing the planet is in better shape than when I found it.

My gratitude and love for you, dear Planet Earth, can be shown with more than words. I will act. In this precious lifetime, I will tread lightly. I will go a step further than carbon neutral and aspire to live carbon positive by taking more carbon from the atmosphere than I use each year. I will aspire to be living without waste, avoiding polluting by reusing and recycling. I will respect the water that I use by consuming what I need and reusing what I can. I will strengthen your exquisite biodiversity around my very own home and beyond. I will take action, one step at a time.

Thank you, Planet Earth. I pledge my love and gratitude to you by making small changes to my life when and where it is possible to do so.

Mia.

GLOBAL MEASURES

Measures vary across Australia, New Zealand, Europe and the US.

LIQUIDS			SOLIDS	
cup	**metric**	**imperial**	**metric**	**imperial**
1/4 cup	60ml	2 fl oz	60g	2 oz
1/3 cup	80ml	2 1/2 fl oz	125g	4 oz
1/2 cup	125ml	4 fl oz	250g	8 oz
1 cup	250ml	8 fl oz	500g	16 oz (1 lb)

OVEN TEMPERATURE			
celcius	**fahrenheit**	**celcius**	**gas**
120 degrees C	200 degrees F	110 degrees C	1/4
160 degrees C	325 degrees F	150 degrees C	2
180 degrees C	350 degrees F	180 degrees C	4
200 degrees C	400 degrees F	200 degrees C	6
220 degrees C	425 degrees F	230 degrees C	8

RESOURCES

Carbon positive home

United Nations Climate Change Lifestyle Calculator: https://www.lifestylecalculator.com/unfccc

Global Footprint Calculator: https://www.wwf.org.au/get-involved/change-the-way-you-live/ecological-footprint-calculator#gs.6ri5p6

Carbon Positive Australia's Calculator: https://carbonpositiveaustralia.org.au/calculate/

Summer and Winter-Proof Home

Green it yourself by Lish Fejer: https://www.greenityourself.com.au

Community Micro forest

The Climate Factory: https://climatefactory.com.au/workshops/

Zero Waste Journey

World Bank Report on global waste to 2050: https://openknowledge.worldbank.org/handle/10986/30317

Shopping: Buy Nothing New and Sharing For Good

Buy nothing new movement: https://www.buynothingnew.com

Gardens big, small and rambling

Gardening Australia: https://www.abc.net.au/gardening/

THANK YOU

An enormous thank you to Amanda and Bea from the HerCanberra team for publishing my sustainable living articles, many of which have been transformed into segments of this book. Thanks to Nikki and Serina who helped and encouraged me to navigate the journey towards publishing.

Thank you to Jess, Michael, Cassie and the Wilkinsons publishing team for beliving in my vision for this book, at every step of the way. Thanks to Dion and Nick for taking gorgous photos and making my home, garden and cooking look increadible. I wrote an initial draft of the book on a little island off the coast of Sumatra, Indonesia. My thanks to Luke and Jane for hosting us there and to grandma Judy for taking care of the kids. Thanks to friends and family who read my early drafts and provided both a critical eye and encouragement, especially to Julie, Jess, Amar, Mike, Eve, Mum and Dad.

Deepest gratitude to Matt, for believing in me and supporting me to live my dreams.

INDEX